URBAN PLANNING AND THE BRITISH NEW RIGHT

Nearly twenty years after Mrs Thatcher first came to power the debate over the significance of what has been termed the New Right continues. Did the 1980s and 1990s see the death of planning and other areas of the public sector as some have claimed or were reports of their demise premature? To what extent were there changes in approach and style as the 1980s progressed and John Major came to power?

Urban Planning and the British New Right attempts to tackle these and other questions through a detailed examination of the 1980s and 1990s. Early attempts to generalise about the implications of a New Right government for planning focused on high-profile policy initiatives such as Enterprise Zones or Urban Development Corporations which led to the conclusion that the 1980s saw the death of planning as a result of the implementation of a New Right philosophy. In this book, leading experts in a wide range of land-use policy areas examine the changes that were brought about during the 1980s and 1990s in planning and the environment, and argue that much less was achieved than expected.

There is little doubt that New Right policy has had a major influence on the shape and direction of British politics. The planning system reacted to these changes in a way that altered or diluted centrally directed proposals. How this came about tells us as much about the shape and power of local planning in the UK as it does about the ability of central government to impose its preferred policies. *Urban Policy and the British New Right* says as much about the administration, institutions and processes of planning as it does about Mrs Thatcher's attempts to change it.

Philip Allmendinger is a Senior Lecturer in the School of the Built Environment, Leeds Metropolitan University and *Huw Thomas* is a Senior Lecturer in the Department of City and Regional Planning, University of Wales, Cardiff.

URBAN PLANNING AND THE BRITISH NEW RIGHT

Edited by
Philip Allmendinger and
Huw Thomas

London and New York

First published 1998
by Routledge
11 New Fetter Lane, London EC4P 4EE

Simultaneously published in the USA and Canada
by Routledge
29 West 35th Street, New York, NY 10001

Typeset in Garamond by J&L Composition Ltd, Filey, North Yorkshire

Printed and bound in Great Britain by
MPG Books Ltd, Bodmin

British Library Cataloguing in Publication Data
A catalogue record for this book is available from the British Library

Library of Congress Cataloguing in Publication Data
Urban planning and the British New Right/edited by Philip Allmendinger
and Huw Thomas.
p. cm.
Includes bibliographical references (p.) and index.
1. City planning—Political aspects—Great Britain. 2. Urban
policy—Great Britain. 3. Environmental policy—Great Britain
4. Conservatism—Great Britain. 5. Great Britain—Politics and
government—1979– I. Allmendinger, Phillip,
1958– . II. Thomas, Huw, 1954–
HT169.G7U695 1998
307.76′0941–dc21 97–39629

ISBN 0–415–15462–6 (hbk)
ISBN 0–415–15463–4 (pbk)

CONTENTS

TABLES

FIGURES

CONTRIBUTORS

Dr Philip Allmendinger is Senior Lecturer in the School of the Built Environment, Leeds Metropolitan University.

Dr Heather Barrett is Senior Lecturer in Geography, Worcester College of Higher Education.

Dr Kevin Bishop is Senior Lecturer in the Department of City and Regional Planning, University of Wales, Cardiff.

Professor Glen Bramley is Chair of Planning and Housing at the School of Planning and Housing, Edinburgh College of Art/Heriot-Watt University.

Christine Lambert is Reader in the Faculty of the Built Environment, University of the West of England.

Dr Peter J. Larkham is Senior Lecturer in the School of Planning, University of Central England.

Neil Harris is a PhD student in the Department of City and Regional Planning, University of Wales, Cardiff.

Angela Hull is Lecturer in the Department of Town and Country Planning, University of Newcastle.

Mark Tewdwr-Jones is Lecturer in the Department of City and Regional Planning, University of Wales, Cardiff.

Huw Thomas is Senior Lecturer in the Department of City and Regional Planning, University of Wales, Cardiff.

Dr Andy Thornley is Senior Lecturer in the Department of Geography, London School of Economics.

Geoff Vigar is Lecturer in the Department of Town and Country Planning, University of Newcastle.

Elizabeth Wilson is Senior Lecturer in the School of Planning, Oxford Brookes University.

ABBREVIATIONS

ADAS	Agricultural Department Advisory Service
CPRE	Campaign for the Protection of Rural England
DNH	Department of Natural Heritage
DoE	Department of the Environment
DoT	Department of Transport
EH	English Heritage
EIP	Examination in Public
EZ	Enterprise zone
LP	Local plan
LPA	Local planning authority
PPG	Planning policy guidance note
RTPI	Royal Town Planning Institute
SoS	Secretary of State
SP	Structure plan
SPZ	Simplified planning zone

1

PLANNING AND THE NEW RIGHT

Philip Allmendinger and Huw Thomas

Introduction

Eight years after Mrs Thatcher's downfall and the advent of John Major's very different style the debate over the significance of the New Right for planning continues. Many may claim that we have moved on – New Labour is now making its own distinctive contribution. But the impact of the New Right has more than simply historical significance. For nearly twenty years planning and other areas of public policy were subject to a distinctive approach regardless of any interpretation of the significance of change. There is little doubt that New Right policy has had a major influence on the shape and direction of British politics and the trajectory of change is of interest and importance to all those involved. At another level the planning system reacted to these changes in a way that altered or diluted centrally directed proposals. How this came about tells us as much about the shape and power of local planning in the UK as it does about the ability of central government to impose its preferred policies. This is a book that says as much about the administration, institutions and processes of planning as it does about Mrs Thatcher's attempts to change it.

In relation to Thatcherite changes Thornley (1993) has identified three distinct perspectives. The 'continuity' view believed that much of the ideological rhetoric would be abandoned when the government faced up to the problems of implementation (Healey 1983). The 'consolidation' view accepts the changes introduced by the Thatcher administrations but believes that their significance can be overstated by ignoring continuities over the post-war period as a whole (Griffiths 1986; Reade 1987). Finally, there are those who believe that the Thatcher years amounted to fundamental change with a move towards a greater reliance upon the market, centralisation of control and minimisation of discretion (Thornley 1993; Ambrose 1986; McAuslan 1980, 1982). Continuity of policy between Thatcher and Major seems largely to have been taken for granted in the planning field. Some have claimed that peace replaced turmoil under John Major (Willets 1992). Thornley (1993) believes that the onslaught was relaxed and that planners had a new lease of life while Rydin (1993) sees a continuation of Thatcherism within John Major's approach. Others have looked in detail at some of the Majorite changes and claimed to identify a hidden agenda in what appeared to be pro-planning changes. The effect of the 'plan-led' system has been to allow central government to dictate local policy through the explosion in policy guidance (MacGregor and Ross 1995). It may be that such implications are 'hidden' because of the relief felt by planners that they were once more seen as 'necessary'. It may be that planning was 'shell shocked', lacking in any alternatives and unclear about what it all means taken as a whole – a postmodern nightmare of too much choice? What is clear is that there is still very little agreement on the significance of what happened in the period 1979–97.

Much of the confusion about the significance of change can be traced back to the nature of analysis. Most studies of planning during this period have concentrated on the link between the philosophy of the New Right and broad changes in the nature of planning (e.g. Ravetz 1980; Healey 1983; Ambrose 1986; Punter 1986; Montgomery and Thornley 1990; Thornley 1993). Considerable weight has been placed on high-profile central government policy initiatives such as Enterprise Zones (Anderson 1983) or Urban Development Corporations (Imrie and Thomas 1993; Brownill 1990) leading, often, to the conclusion that the 1980s saw the 'death of planning' as the result of the implementation of a 'New Right' or Thatcherite philosophy. As Marsh and Rhodes (1992) point out in their assessment of wider Thatcherite changes, many of these assessments were not based upon any thorough analysis of the content or effect of policy. Consequently, such approaches can be questioned on two grounds. First, they assume an automatic transmission of government policy into local practice, an assumption which developments in implementation theory have demonstrated to be simplistic and misleading (Barret and Fudge 1981). Second, grand overviews of planning under the New Right have ignored the significance of recent theorising about the importance of the uneven spatial development of capitalism, including the uneven development of local social relations (e.g. Bagguley *et al.* 1990). Even an account such as Brindley *et al.* (1996) which acknowledged the variety of local planning practices in the 1980s portrayed it as a response to contemporary economic conditions, and did not really explore the significance of variations in local social relations in shaping responses. Healey *et al.*'s (1988) analysis of planning *did* appreciate the need to account for the 'locality effect', but focused largely on only one aspect of planning, namely development planning.

Early attempts to generalise about the implications of a New Right central government for planning were handicapped by the lack of detailed studies of planning practice, and so it is perhaps inevitable that they fell foul of the criticisms set out above. However, detailed studies of many aspects of the planning system, and a number of policy initiatives, are now available, and so a more sophisticated picture can be assembled. This book undertakes that task by bringing together a number of studies on aspects of planning practice in the 1980s and 1990s. Using a conceptual framework based on implementation theory and an appreciation of the significance of locality the book's distinctive contribution to this debate is detailed case studies of changes to local planning practice. Through this approach it hopes to provide an authoritative account of the trajectory of planning today, as well as a critique of accounts which have relied too heavily on generalisations and policy intentions rather than research into local practice.

The New Right

Before embarking on a study of local planning practice under the New Right we need to address a number of fundamental questions: Is there such a thing as the New Right? Was there a distinctly New Right approach to planning? Are there enough similarities between Thatcher and Major to allow us to label both New Right? We believe the answer to all of these questions is 'yes', but a qualified 'yes' to the third question.

Though some, such as Riddell (1983), Jenkin (1984), Hirst (1989) and Bulpitt (1986) question the consistency of Thatcherism the majority view is that Mrs Thatcher had a coherent set of political ideas and these guided her behaviour. Such ideas were based around how the economy should be organised and the style and content of government (Thornley 1993). These two strands have been variously labelled social market economy and authoritarian popularism (Gamble 1984), free economy and strong state (Gamble 1988), economic liberalism and authoritarianism (Edgar 1983), neo-liberalism and combative Toryism (Norton and Aughey 1981), and liberalism and Conservatism (King 1987). However these two strands are labelled all of the authors point to the move in Britain since 1979 towards a freer, more competitive, more open economy and a more repressive, more authoritarian (and centralised) state (Gamble 1984: 8). Various studies have shown that Thatcher's commitment was translated into policy (Marsh and Rhodes 1992: Kavanagh and Seldon 1989), though what has been termed 'electoral popularity' (Hirst 1989) or 'statecraft' (Bulpitt 1986) also had a significant impact upon policy.

While the Thatcher agenda appeared to be clear one searches in vain for a similar expression of John Major's beliefs (Kavanagh 1994). According to Kenneth Clarke, John Major received support from colleagues because he was 'Mrs Thatcher with a human face' *and* because he represented a break with the past (Crewe 1994). This confusion (deliberate or not) dogged the Major governments. Nevertheless, there is little doubt that *some* of the Thatcherite agenda continued – proposals to privatise the Post Office, a monetarist emphasis on controlling inflation through the money supply, deregulation, market testing in Whitehall, etc. But there was sniping from the Right over tax rises and Europe. Most tellingly, John Major emphasised the community while true Thatcherites followed the 'no such thing as community' line. A number of factors help to explain this confusion and the forces pulling both away from the Thatcher legacy and back towards it. Some are personal, others contextual. First, let us consider those pushing for change.

Conciliation rather than confrontation

Mr Major's background as a whip taught him to conciliate rather than confront and this has meant that his own agenda was less obvious than his predecessor (Riddell 1991). His small majority in the Commons, his party's disunity over the future of the European Union, disaffected and disenchanted MPs, and withdrawal from the Exchange Rate Mechanism (ERM), all required these skills of brokerage whereas Mrs Thatcher could use her large majorities to force change (Marsh and Rhodes 1992).

The loss of common 'enemies'

Mrs Thatcher had had a large number of enemies against which she could unite the party. The unions, nationalised industries, Europe, the civil service, inflation, etc. all allowed and required strong conviction leadership. By 1990, as Kavanagh (1994) points out, it was no longer clear who the enemies were and the 'easier' targets (e.g. unions) had mostly gone. Mr Major inherited more complicated issues – you could not simply privatise inner cities, though the Thatcher governments did attempt to promote a market-led solution to perceived inner-city problems by setting up Urban Development Corporations.

From economic to social

Mrs Thatcher had spent most of the 1980s concentrating on economic issues. By 1990, by-election results and opinion polls were pointing to a switch in emphasis to social problems – e.g. rising crime, the breakdown of community, the future of the National Health Service. Again, it became clear that these issues could not be dealt with simply by invoking the social policy equivalent of 'low inflation' or 'money supply' as the failure of 'Back to Basics' and other slogans demonstrated. Even Conservatives concede that some might require distinctly different non-Thatcherite approaches.

External pressure

Too much emphasis may have been put upon the uniqueness of Thatcherism and especially its approach to economic matters. Marsh and Rhodes (1992) point to many social and economic changes that dictated policy during the 1980s (demographic change, deindustrialisation, rising unemployment, etc.) and it is true that most of the policies pursued by the Thatcher governments were also pursued by other governments around the world including governments of the Left in countries such as France, Spain, Australia and New Zealand (Kavanagh 1987). It is also the case that macro-economic policy is increasingly dictated by the world money markets and that the European Commission has significant influences upon policy notwithstanding the Social Chapter opt-out under both Major and Thatcher. These pressures have *increased* since 1990 and the scope for unilateral policy formulation has accordingly diminished (Kavanagh 1994).

The above were pressures upon the Major government to modify the Thatcherite approach. There were also strong influences for continuing the Thatcher agenda.

Public support

Regardless of the evidence of polls that the public felt the need to pay more attention to the neglected social aspects of policy there was equally no inclination (with the exception of the poll tax) to throw away the main planks of Thatcherism. On the contrary, there was widespread public support for most of the policies pursued during the 1980s (especially law and order, union restraint and privatisation) (Kavanagh and Seldon 1994).

The middle ground

The policy debate throughout the 1980s had shifted, and it is often claimed that Mrs Thatcher had as much impact on the Labour Party as on the Conservatives (Young 1994). By 1990 Labour had moved from being a party of socialism to one resembling European Social Democracy (Marr 1995). The common ground between the two main parties now includes:

- a narrowing of choice in economic policy through the EU, the discipline of international money markets and the lack of support for tax rises
- an acceptance of the need to make public services more responsive and efficient

- the need to rebuild a sense of 'community'
- a greater emphasis on the use of markets to allocate resources, especially in the public sector (Kavanagh 1994).

This leaves little scope for innovation and goes some way towards what Geoffrey Howe once called 'winning the debate' (Anderson 1983). The defection of Alan Howarth to Labour in 1994 and Emma Nicholson to the Liberal Democrats in 1995 has demonstrated that winning such a debate is not all to the Conservatives' benefit.

All of the above factors clouded the direction of government between 1990 and 1997, and could fairly be summarised as pointing towards and away from a continuation of the Thatcherite approach. In terms of *policy* Kavanagh and Seldon (1994) conclude in their wide-ranging review that, on the whole, there *was* a continuation of Thatcherism under John Major though he put his own stamp on policy. Some of the changes in urban policy illustrate this complexity. Originally set up under section ix of the Local Government, Planning and Land Act 1980, Urban Development Corporations (UDCs) were, for some time, seen as flagships of the Thatcherite approach to urban renewal, which (in the words of a leading civil servant) 'has an economic focus, concentrating on supply side measures, with the leading role for the private sector' (Solesbury 1990, quoted in Atkinson and Moon 1994: 97).

UDCs promoted property-led regeneration (Imrie and Thomas 1993). Their powers (notably of compulsory purchase, and – in most areas – planning), and the measures against which their performance was measured, were intended to focus their attention and energies on creating attractive conditions for private-sector investment. In time the economic benefits of such investment would 'trickle down' to the population as a whole (for example in the form of jobs in the offices, factories or shops which had been developed), though in the short term those with interests in property and construction would be clear gainers from UDC activity. Dependent as they were (and are) on the private sector to implement the bulk of their objectives UDCs have tended to eschew major planning exercises or strategy formulation. Most of them produced 'strategies' or 'plans' in their early days which smacked more of marketing than of planning: a broad set of objectives, or a 'mission', and a series of photographs and artistic perspectives. The documents are clear evidence of the priority attached to attracting and sustaining investors' interests in their areas. Even Cardiff Bay Development Corporation (CBDC) which – unusually – undertook a year-long strategic planning process has not bothered to publicly update its plan, now over ten years old – meanwhile, it has responded on an *ad hoc* basis to investment interest from the public and private sectors. The development corporations have been insulated from their localities because their boards are appointees of the relevant Secretary of State (of the Environment, or of Wales). Even prominent local councillors who have a place on a board are there as nominees, not as representatives of their local authorities. Not surprisingly, UDCs have paid relatively little attention to community relations and local people have found them difficult to influence (Brownill and Thomas 1997).

City Challenge, set up in 1991, and Single Regeneration Budgets (SRB) (instituted in 1994) provide a contrast to UDCs which it is persuasive to regard as a Majorite twist in urban policy. Though both have a strong emphasis on economic regeneration, this is by no means as pronounced as within UDCs, where typically up to 3 per cent of budgets have been allocated to community development (Imrie and Thomas 1993: 16). Moreover, whereas the economic focus of UDCs more or less ensured that a very narrow band of property-related interests were the major beneficiaries in the first instance, the economic orientation of City Challenge and SRBs has allowed for schemes (such as training and education) where the direct beneficiaries may be on low incomes. Another contrast between UDCs and the newer agencies is that while the former have been created by central government, which appoints their boards, the latter are, to some extent, local creations, 'partnerships' (so-called) of local authorities, the community (generally, in the form of voluntary organisations) and the private sector. City Challenge and SRB agencies rely on central government funding, for which they bid; the rules of the 'bidding game' (including the need to demonstrate 'partnership') are set down (both formally and informally) by central government. This is a major constraint (Oatley and Lambert 1995), but the precise composition of the City Challenge or SRB 'team', and the strategy it devises is ultimately a local matter. Note that there *is* a strategy. Competitive bidding demands plans – costed, with output targets and 'milestones' – which can be compared (and reviewed). In addition, important as central government guidelines and policy undoubtedly are, there can be no doubting that City Challenge and SRBs have the scope, at least, to be sensitive to local communities and circumstances in ways which UDCs have always found difficult. One interpretation of this shift is that urban policy has tentatively begun to acknowledge 'society', or community, once again.

Yet, different as City Challenge and SRB are from UDCs, continuities must also be acknowledged. Foremost among these is their role in attempting to check the importance of elected local government in formulating and implementing public policy at the local level. A variety of pressures have encouraged and pressured local authorities to change their mode of operation, and their remits, and among the most significant has been the Conservative project of curbing elected local government as both a primer of public expenditure and a focus for political opposition. UDCs are a spectacular example of an attempt to elbow aside elected local authorities in a policy area where they had a well-developed role (and expertise); in City Challenge and SRB, local authorities are at the heart of partnership-creation, but every effort is made to ensure that the agencies they help create are not their creatures. Another significant continuity is that both UDCs and City Challenge are fixed-term 'solutions' to urban problems. In part this stems from Conservative fears of creating self-perpetuating bureaucracies of the kind many feel local authorities to be; but it also betokens a feeling that the task is to re-integrate problematical places into the mainstream of the urban economy and that once this is done the state's role will have been fulfilled.

This picture of continuity and change based on central government's policy aspirations and legislation becomes more complex still when we examine the realities of policy delivery on the ground. On the one hand, it is clear that the policies and approaches of individual UDCs have varied over time and between places. Brownill (1990, 1993) analyses the way in which the London Docklands Development Corporation (LDDC) became at one period more ready to tailor policies and projects to deliver tangible short-term gains to its local population. This sensitivity to local and national political pressure does not appear to have been sustained into the mid-1990s, but it does illustrate the dangers of too simplistic a view of the UDC or City Challenge 'approach'. Every UDC has had to create working relationships with a range of local agencies (including local authorities), and in some areas these have developed into close ties (Thomas and Imrie 1997; Imrie and Thomas 1993). Some UDCs have accepted, and worked within, politically charged policies of local authorities, such as equal opportunities policies; others have guaranteed minimum levels of social housing provision as part of their strategies (Dabinett and Ramsden 1993; Thomas and Imrie 1993). Conversely, whatever the rhetoric of 'partnership' in City Challenge projects, available evidence suggests that community groups remain marginal in many, unable to exercise significant influence in policy discussions couched in terms more familiar to bureaucrats and business people (Davoudi and Healey 1995; Macfarlane 1993). In some areas it is clear that partnership must be largely on the private sector's terms (Oatley and Lambert 1995).

On balance, it would be difficult to deny that there have not been important changes in urban policy under the New Right. Some appear to be consistent with the more conciliatory, community-oriented version of New Right policy associated with John Major. But it would be mistaken to read the changes as simply a translation into policy of a new(ish) political view, for the recession in the property market in the late 1980s exposed the narrowness of approach of UDCs, while business involvement in Training and Education Councils (TECs) and other agencies of local governance began to defuse the political suspicion that any kind of involvement with local authorities was tainted. What we now need to assess is whether the same complex process of change extended to land-use planning and what change in direction, if any, there was.

From Thatcher to Major

The 'plan-led system' introduced in 1991 has led the renaissance of planning though initial inflated claims (e.g. Edwards and MacCafferty 1992; Taussik 1992) have been followed by more sober reflection (e.g. MacGregor and Ross 1995). Some have maintained that the plan-led approach can be traced back to a reaction against the more *laissez-faire* approach of the 1980s (Thornley 1993). Others (e.g. Healey 1992) point to the government's need to accommodate its commitment to sustainable development made at the Rio Earth Summit. But suggestions that the New Right suddenly 're-discovered' planning and unilaterally decided to shift the emphasis from market to plan may be naive. Hirst (1989) has claimed that the Conservative Party is an umbrella organisation of different interests and these interests may often conflict with each other, e.g. the conservationary-minded shire voter and the property developer (Elson 1986). The *laissez-faire* approach of the 1980s favoured the latter interest though property recession led some to believe that the market needed to be protected from itself – and planning achieved this. Similarly, the 'green backlash' added to the government's need to maintain support in an important part of the electorate (McCormick 1991). Greater understanding of the direction of change is also now emerging. The move towards a plan-led system has involved an explosion in central policy guidance (Tewdwr-Jones 1994), accusations of increased central control in local matters and long delays in preparation of plans. Rydin (1993) and Thornley (1993) have also claimed that the replacement of the presumption in favour of development with the presumption in favour of the development plan effectively shifts the emphasis away from the market. However, the idea of a shift from the market to the plan misses the point that the presumption in favour of the development plan sits alongside the presumption in favour of development (MacGregor and Ross 1995). There seems to be no problem in applying the presumption in favour of development where the development plan contains a presumption against the proposal. The balance between the two has to be interpreted by the courts who have regard to policy advice. Such advice in the 1990s has retained a strong presumption in favour of development (see Planning Policy Guidance Notes PPG 1, PPG 6 para. 3.1, PPG 4 para. 13). Even attempts at demand management in PPG 13 have been criticised for being half-hearted and likely to have been achieved in any event (Headicar 1995). While some of the more strident market-oriented approaches have been diluted (e.g. in relation to design) the market legacy has continued into the 1990s, though in a less obvious form.

What we can conclude is that although John Major's government had a very different style from Mrs Thatcher's and in some ways pursued different policies the underlying themes of the 1980s have remained. Planning has not been exempt from New Right influence. While some question consistency few would disagree with Thornley's broad conclusion that 'The intention of government is to retain the bones of the planning system but give it new shape. The purpose is one which has its primary aim in aiding the market' (1991: 143).

Much of the assessment of planning under the New Right has assumed an automatic transmission of policy (e.g. Thornley 1993; Ambrose 1986, 1992; Ravetz 1980; Healey 1983), while the significance of localities and distinctive policy processes has been largely overlooked. Some works have questioned this assumption. One of the first studies to focus on local change and question the overriding dominance of Thatcherism was Brindley *et al.* (1996). Although acknowledging the variety of planning practice in the 1980s they portrayed it as a response to contemporary economic conditions. Brownill's (1990) study of the London Docklands Development Corporation emphasised the influence of social *and* economic factors. Docklands has a long history of local action and involvement and throughout the 1980s community campaigns pushed for alternatives to the Corporation based on the needs and experiences of local people. Criticism of the Corporation was levelled by Parliament and some developers themselves at its 'single minded' and 'top-down' approach and the lack of attention to views and needs of local residents (Brownill 1990: 5). Such local influences led to a dilution in the Thatcherite approach: 'While inner city leverage may have gained the ascendancy, tensions and conflicts within it as a policy intervention have led to compromises and changes over time, as has continued pressure from the locality' (ibid: 172).

The homogeneous view of central initiatives such as Urban Development Corporations was also questioned by Imrie and Thomas (1993) who argue that much research on the subject has tended to take a reductionist line in portraying them as invariant, undifferentiated central initiatives. In reality, the work of Development Corporations requires forging links with local politicians, community groups and local civil servants. Following the London Docklands experience, later Corporations attempted a softer, more conciliatory approach to local consultation by devising closer links with a range of community and local organisations. Both Brownill (1990) and Imrie and Thomas (1993) reflect on the legacies of socio-economic and political backgrounds of different localities which were fundamental in shaping the precise configuration of (local) policy content and implementation. This is also the focus of Healey *et al.* (1988) who examine the ways in which the planning system has been adapted to local circumstances in recent years. They too acknowledge the influence of localities upon policy formulation and implementation and start from the premise that every locality is a unique configuration of economic activities, divisions of labour, cultural traditions, political alignments, spatial arrangements and physical form. They conclude that not only is there a 'locality effect' in terms of a locale's response to and interaction with policy but also in terms of interests, including those that the planning system favours: 'Exactly what is in the interests of particular farmers, mineral operators, developers and investors in a particular place and time is likely to be highly specific to the location and the particularities of the firms' strategies' (p. 246).

This 'multi-layering' of the significance of localities opens up another aspect of the debate concerning the significance of Thatcherism. It was recognised by some (e.g. Griffiths 1986; Hirst 1989) that the Conservative Party is a 'broad church' of interests and needs to pursue electorally popular policies that have the support of its 'natural' allies. Various studies have pointed to the government watering down policy or abandoning it altogether when unfavourable responses from these interests were received (Thornley 1993; Elson 1986). According to some this eventually led to the shift from the project-led system of the 1980s towards the plan-led system of the 1990s (MacGregor and Ross 1995; Healey 1992; Thornley 1993).

There are precious few studies of planning at the local level during the 1980s and 1990s. Nevertheless, the above studies raise a question mark over the deterministic studies of government policy during this time and point to the significance of localities in the implementation of policy. In order to provide a conceptual background to the study of local planning practice under the New Right the next section provides a brief overview of the literature on implementation and localities.

An implementation perspective

The idea that an implementation perspective can provide a useful insight into the extent of change during the Thatcher years is not new (Marsh and Rhodes 1992), though this approach is yet to be applied to planning or indeed to local government. A study of the impact of the New Right agenda requires a review of existing theories of implementation. As Ham and Hill (1993) and Marsh and Rhodes (1992) point out, there are two basic approaches to the study of implementation. The 'top-down' approach (characterised by the separation of implementation and policy-making processes) which developed from studies in the United States during the 1970s (Pressman and Wildavsky 1984) has been widely criticised especially in relation to experience in the UK. An alternative 'bottom-up' approach has been preferred that treats implementation as a political rather than managerial problem (Barret and Fudge 1981; Elmore 1982; Sabatier 1986).

In their influential book, Pressman and Wildavsky (1984) define implementation as 'a process of interaction between the setting of goals and achieving them (1984: xxiii). As Ham and Hill (1993) point out, the starting point in this approach is the identification of policy and involves a distinction between policy-making and implementation. This approach and others that take a similar line (e.g. van Meter and van Horn 1975) stress the importance in successful implementation of linkages between organisations and departments at the local level. If these links are not close then 'implementation deficit' may occur. The transfer of this concept to the study of the UK administrative system was undertaken by Hood (1976) who discusses the 'limits to administration', concentrating not so much on the political processes that occur in administration but on controls that limit complex administrative systems. Further work by Dunsire (1978) develops these ideas into an abstract model of problems to be faced by persons attempting 'top-down' control over the administrative system. The tendency to prescribe preconditions for successful implementation is characteristic of 'top-down' approaches. Hogwood and Gunn (1984) set out ten such preconditions:

- that circumstances external to the implementing agency do not impose crippling constraints
- that adequate time and sufficient resources are made available for the programme
- that not only are there no constraints in terms of overall resources but also that, at each stage in the process, the required combination of resources is actually available
- that the policy to be implemented is based on a valid theory of cause and effect
- that the relationship between cause and effect is direct and there are few, if any, intervening links
- that there is a single implementing agency which need not depend upon other agencies for success or, if other agencies must be involved, that the dependency relationships are minimal in number and importance
- that there is a complete understanding of, and agreement upon, the objectives to be achieved; and that these conditions persist throughout the implementation process
- that in moving towards agreed objectives it is possible to specify, in complete detail and perfect sequence, the tasks to be performed by each participant
- that there is perfect communication among, and co-ordination of, the various elements in the programme
- that those in the authority can demand and obtain perfect obedience.

Ham and Hill (1993) conclude that such an approach seeks to provide advice to those at the top on how to minimise implementation deficit. Sabatier and Mazmanian (1979) have four preconditions of their own:

- ensure that policy is unambiguous
- keep links in the implementation chain to a minimum
- prevent outside interference
- control implementing actors.

Criticism of the 'top-down' approach has been on a variety of grounds. First, too much attention is seen to be given to the objectives and strategies of central actors and too little emphasis on the role of other actors in the process (Lipsky 1978). Second, the conditions for successful implementation are seen by some as unrealistic – there is always a scarcity of resources (Barret and Fudge 1981). Third, discretion is inevitable in all organisations – the activities of 'street-level bureaucrats' will lead to implementation deficit (Elmore 1982). Fourth, the 'top-down' approach focuses on the identification of policy and therefore ignores the unintended consequences of government action (Hjern and Hull 1982). Fifth, some policies do not have, nor were they intended to have, explicit objectives – they grow and evolve over time – and therefore lack benchmarks by which to measure them (Hogwood and Gunn 1984). Finally, the theoretical distinction between policy formulation and implementation cannot be sustained in practice because policies are made and remade in the process of implementation (Sabatier 1986; Barret and Fudge 1981).

The thrust of these criticisms has led to a focusing on individual actions and actors who respond to choices or issues (Elmore 1982) and the view of implementation as a policy–action continuum (Barret and Fudge 1981). This approach typifies the 'bottom-up' view of implementation which, as Ham and Hill (1993) point out, is relatively free of the predetermining assumptions of the 'top-down' alternative. Barret and Fudge (1981) and Hjern and Porter (1981) see the basic unit of analysis in implementation as being the service delivery network – implementation is seen as a negotiating process in which individual actors pursue their disparate objectives through multiple strategies (Marsh and Rhodes 1992: 7). As opposed to the benchmark emphasis of policy objectives in the 'top-down' approach the 'bottom-up' variant concentrates on the multiplicity and complexity of linkages, the problems of control and the co-ordination and management of conflict and consensus.

Some common criticisms levelled at the 'bottom-up' approach include its emphasis on the discretion available to street-level bureaucrats who are in fact subject to legal, financial and organisational constraints (Marsh and Rhodes 1992). Although such parameters do not determine behaviour they set parameters upon discretion. Second, factors that influence actors' perceptions, views and actions are not explained. The origins of the bureaucratic processes which frame the influence of actors as well as the distribution of resources between actors is crucial to the 'bottom-up' approach though it is not explained in detail by the theory. Sabatier

(1986) also points out that the proponents of this approach are not concerned with implementation *per se* but with understanding the actions and interactions of a policy process. Finally, this approach excludes circumstances where policy objectives are made explicit and do structure the decision-making environment of local actors.

As a great deal of policy is actually made, or modified, in the implementation process it follows that concern about the impact of officials or bureaucracies must extend to a larger group than the top echelons concentrated on in the 'top-down' approach. Officials, or bureaucrats, in the public sector have a number of distinct characteristics that can lead to the modification of policy. Merton (1957) argues that bureaucrats are likely to show particular attachment to rules that protect the internal system of social relations and enhance their status by enabling them to take the status of the organisation to protect them from conflict with clients by emphasising impersonality. A number of reasons are put forward to explain this including the role of public scrutiny which emphasises conformity with rules and job selection and career promotion that stresses a regularised career structure and conformity with the organisation. Thus bureaucrats become advocates and bargain and negotiate on behalf of their organisation with other 'policy sectors' (Benson 1983).

The drawbacks of this analysis have been highlighted by Lipsky (1978) who focuses on the discretion available to bureaucrats who can (and do) make choices to enforce some rules, particularly those which protect them, while disregarding others. This is particularly true of professionals within organisations (Wilensky 1964). According to Ham and Hill (1993) professionals have succeeded in persuading politicians and administrators that the public sector will receive the best service if their discretionary freedom is maximised. What this adds up to, according to Simon (1945), is that within an organisational system a series of areas of discretion are created in which individuals have freedom to interpret their tasks within general frameworks provided by their superiors. Dunsire (1978) sees this as creating 'programmes within programmes' where subordinate programmes are dependent upon superior ones but may involve different kinds of activities. In fact, in a hierarchical situation superiors may be dependent upon subordinates to implement policy.

The distance between these two views is not as clear as others (e.g. Marsh and Rhodes 1992) have argued. As with the difference between the 'top-down' and 'bottom-up' views of implementation the bureaucratic conformity and discretion views can exist side by side in an organisation. The relationship between organisations and the concept of 'policy sectors' has been developed to examine the links between local government and other organisations and bodies including quangos. Boddy (1983) has pointed to the increasing complexity of state agencies and conflicts that arise; these have led to local agencies receiving confused signals from the centre and to problems of accountability. According to Cloke (1986) this situation has led to an interest in research into networks. Empirical work on this approach has covered many areas of public policy and Smith (1990) demonstrates how one such network existed from 1945 to the early 1980s concerning agricultural policy with two dominant actors, the Ministry of Agriculture, Fisheries and Food and the National Farmers Union. Marsh and Rhodes (1992) believe that networks affect policy outcomes and constrain the policy agenda. Further, there is ample evidence from studies that policy networks allow for policy continuity and are effective at resisting change – 'dynamic conservatism' as Rhodes (1992) terms it. These networks are normally a two-way conduit of influence that feeds into policy formulation and is instrumental in implementing it. The Thatcherite approach to government during the 1980s rejected the use of networks (Rhodes 1992: 73) and thereby took a 'top-down' view of policy implementation that alienated those implementing it (Marsh and Rhodes 1992).

The above review is by no means exhaustive (for example it only touches on important areas such as the role of bureaucracies, bureaucrats, professional organisations, discretion and political pressure). What it does is demonstrate two important points. First, it is naïve to expect an automatic transmission of policy into practice and, more radically, that the very terms with their implications of dichotomy need to be re-thought. Second, that there is bureaucratic discretion within and between organisations that provides scope for differences in policy implementation. One of the key determinants of how this potential for variation is realised is the nature of local social relations, of 'localities'.

The role of localities

The starting point for the analysis of uneven spatial development is 'the geographically uneven spread of those factors which affect the profitability of production processes' (Bagguley *et al.* 1990: 15). Historically, Duncan and Goodwin (1988) suggest, these may have been predominantly natural resources, the uneven distribution of which will have made production of certain goods more profitable in one location than another. However, in Bagguley *et al.*'s (1990: 150) formulation, 'New rounds of investment will be geographically patterned in response to the pre-existing spatial pattern' – i.e. in searching out opportunities for profitable investment capitalists will, perforce, take into account the package of resources and constraints offered by particular geographical locations, and these will include not only natural resources (such as mineral deposits) but also the kind of built environment which has been developed (including transport and other infrastructure) and, increasingly, 'human resources' (the skills, aptitudes and attitudes of the available workforce). There is, of course no guarantee that places suitable for a certain kind of production will remain the most profitable location area for that kind, should technologies change (and technological change, for our purposes, can involve not simply advances in science-based technology, but also new methods of management or workplace organisation). But moving locations of production is not without costs as Harvey (1989) reminds us, and these can sometimes be high (e.g. for firms engaged in mass production of consumer goods).

This perspective on the spatial dimension of capital accumulation provides a basis for understanding how different locations will contain different mixes and patterns of buildings: some of those differences, at least, will relate to different opportunities – and increasingly different opportunities – for profitable investment. However, production involves social relations as well as technologies – factories or offices are organised in particular ways and are staffed by particular groups of people. Moreover, the production process does not exist in a social or economic vacuum – the workforce must be fed, housed, entertained, educated and so on; in a word, there is a process of reproduction of the workforce, which will also be organised in a particular way, incorporating technologies and involving particular kinds of social relations. Duncan and Goodwin (1988: 69), in particular, stress the significance of the uneven spatial development of capitalism in creating spatially distinct patterns of social relations in their spheres of production and reproduction, as new rounds of investment interact with existing physical and cultural/social patterns:

> The practices of civil society are constituted contingently, in the context of nature, of each other and of world capitalism. For example, gender divisions of labour in simple gatherer–hunter societies . . . owe much to the cultural interpretation and organisation of labour . . . The same principle holds in capitalist societies, except that now the uneven development of capitalism overlaps natural unevenness.
>
> (Duncan and Goodwin 1988: xx)

Mark-Lawson and Warde (1987) trace examples of the locally specific connections between social relations in the process of production, the domestic sphere and urban politics. Savage's account of weaving in Preston from the late nineteenth century to the inter-war period illustrates the way in which patriarchal structures in the domestic sphere (the sphere of reproduction) extended to and were associated with patriarchy in the workplace. His evidence shows that in pre-war Preston's weaving sheds it was unusual for (male) heads of households whose daughters or wives were employed as weavers to agree variations in working conditions (e.g. time off) directly with their (male) overseer, who – in the weaving shed – acted as a kind of surrogate head of household, regulating morals as well as quality of work. However, in the early twentieth century, though patriarchal relations remained, the role of the overseer in sustaining them was undermined by changes in both the forms of ownership of mills (joint stock companies with specialist managers becoming more usual), and state-sponsored change in the labour market – namely the introduction of National Insurance – which reduced the scope for discretion in recruitment on which overseer power rested in large measure. The case study demonstrates some of the links between the spheres of production and reproduction, and also, as Savage remarks, the intertwining of local conditions, changes in the management of capitalist enterprise and state policies implemented uniformly throughout the country.

Mark-Lawson and Warde (1987) explore the implications of gender relations in the workplace for urban politics, comparing Preston with the neighbouring towns of Lancaster and Nelson. The comparisons between Preston and Nelson, both cotton towns, are especially interesting. They are persuasive in arguing that the general segregation in Preston's labour market (whereby, weaving was largely a female occupation supervised by men, with men also employed in other industries to which women were not recruited) is central to understanding the low priority attached by the local Labour Party to welfare issues. The Preston Labour Party's links to trade unions, from which women were excluded by patriarchal attitudes and power, isolated women from it and also insulated the party from the influence of women activists. In Nelson, on the other hand, employment opportunities were not restricted to cotton, with the result that substantial numbers of men worked alongside women as weavers, both subject to the supervision of (male) overseers and managers. This experience of equality of conditions and lack of segregation in the labour market underpinned women's involvement in both trade unions and labour politics, with a correspondingly greater prominence attached to welfare issues than in Preston.

Bagguley *et al.* (1990: 185) suggest that although there is no direct correlation between economic restructuring and political action these events will shape political action, and previous events set the agenda for change, shaping the issues that will be pursued, the groups involved and the resources available to them. Local government has become a focus for these forces, perhaps, as Bagguley *et al.* (1990) suggest, because of the dominance of Thatcherism during the 1980s. Duncan and Goodwin (1988) explore the role of local government against the backdrop of such local policy processes:

> Because social relations are unevenly developed there is, on the one hand, a need for different policies in different places and, on the other hand, local state institutions to formulate and implement these variable policies. Local state institutions are rooted in the

> heterogeneity of local state relations, where central states have difficulty in dealing with this differentiation. But . . . this development of local states is a double edged sword – for locally constituted groups can then use these institutions to further their own interests, perhaps even in opposition to centrally dominant interests.
>
> (p. 114)

As Bagguley *et al.* (1990) put it:

> The local state then becomes both a means by which central government deals with the problematic effects of uneven development and mode of representation of interest groups potentially opposed to the centre. From the potentially contradictory nexus comes the dynamics of local policy variation.
>
> (p. 185)

Again, this is only a brief review of a very large literature. What it does demonstrate is that there is undoubtedly scope for autonomous local politics and, as Bagguley *et al.* conclude, place does matter. Duncan and Goodwin (1988) argue that the existence of local government is predicated upon a tension inherent in the uneven development of localities and regions in capitalist societies: central authorities attempt to impose uniform regulations on all places, while people in localities demand, given the unevenly developed economy, that specific local interests be effectively promoted. The result is that place is an important (though complex) influence on local politics and political action.

The structure of this book

The book has ten chapters in all. In chapter 2 Elizabeth Wilson examines the role of the environment throughout the 1980s and 1990s in New Right policy, highlighting the contradictions and changes. As the environment moved up the political agenda it introduced a new and more politically charged set of factors that the New Right found difficult to deal with within their ideological framework. Some local authorities grasped the new agenda and its tools, often leading the government into more environmentally sensitive policies.

Another area that has proved a perpetual headache for the New Right has been conservation and design. Peter Larkham and Heather Barrett explore the initial antagonism towards conservation from the Conservative governments as well as the various ways in which it has continued to flourish and remained a popular cause in many localities. In contrast Glen Bramley and Christine Lambert focus on housing in chapter 4 and the almost complete failure of the New Right to achieve its aims of facilitating the role of private development and minimising supply-side constraints such as planning. Nowhere was the 'U' turn on government policy between the 'project-led' approach of the 1980s and the 'plan-led' approach of the 1990s more evident than in changes to development plans. Angela Hull and Geoff Vigar demonstrate in chapter 5 that the plan system is now shifting towards a European prescriptive model with increased central government guidance homogenising policy and restricting locally led responses.

In an assessment of one of the most ideological approaches to planning Philip Allmendinger charts the impact and influence of Simplified Planning Zones and their failure to radically alter the landscape of planning. A combination of unclear policy objectives, inadequate causal theory and local resistance led to a botched attempt to shift the UK system towards a market-led approach.

The significance of locality is also emphasised by Mark Tewdwr-Jones and Neil Harris. In chapter 7 they consider development control under the Thatcher and Major governments. They argue for a distinctive Majorite approach to development control in the 1990s and also identify an attempt to create uniformity in the kinds of criteria used in decision-making on planning applications. They suggest that there is evidence of local resistance to this drive for uniformity, and that the ensuing wrangles between central and local government are often couched as arguments over 'malpractice'.

Whereas central government interest in development control has generally been decisive, it has had to tread more warily in relation to countryside conservation, as Kevin Bishop's chapter makes clear. Both nationally and locally there are complex configurations of political and economic interests surrounding countryside policy, many of these with long-standing ties to the Conservative Party. Bishop reviews a process of steady increases in central government sponsored regulation of the countryside activities accompanied by attempts to introduce elements of the 'enterprise culture'. Meanwhile, many local authorities, reacting to immediate economic distress, have promoted (again, cautiously) economic development policies. This chapter – perhaps more than any other – also points to the growing significance of European policies (and politics) in influencing local responses, for example, in allowing leverage for local dissent from (national) government initiatives.

Andy Thornley reflects on the consistency and significance of eighteen years of New Right government in chapter 9. He comes to the conclusion that despite changes in leader and policy priorities, the Thatcherite concern with markets and centralisation still underpinned the government's approach throughout the 1980s and 1990s. Although, as we claim above, there is scope for autonomous action, Thornley questions the extent to which this allowed local authorities and others to depart from highly centralised approaches.

This view is to some extent a departure from the main argument of this book and is analysed in the final chapter along with the main questions raised earlier. Here we explore the evidence presented in the different chapters and what it says about both the New Right approach to planning and planning itself.

2

PLANNING AND ENVIRONMENTALISM IN THE 1990S

Elizabeth Wilson

Introduction

British land-use planning seems to have displayed a remarkable resurgence of activity at the local level in the 1990s. While this must have surprised those who foretold the death of planning by the end of the 1980s, many observers and practitioners (for instance, Healey and Shaw 1993; Marshall 1994; Owens 1994; Brindley *et al.* 1996) have suggested that it is the grasping of the environment as an issue which has generated this activity, and rescued planning from the intellectual and professional doldrums of the 1980s.

This chapter will examine some of the arguments for this view, and explore the debates over the real extent and possible consequences of any material change in the scope of planning in the 1990s. Using the evidence of activity amongst local planning authorities, other regulatory agencies, and pressure groups with an interest in the planning process, it attempts to explain the wide variety of experience in terms of some of the contradictions inherent in reconciling the adoption of the potentially radical concept of environmental sustainability with the conservative corporatism of both central and local government in the UK.

Exploring the contradictions

These contradictions themselves reflect shifting positions in the broader political context within which planning is undertaken. Partly these derive from the uneasy relations between local government and central government in this period (discussed elsewhere in this volume), and partly from the uncertainty with which central government has grasped the environmental nettle – there is still a very real debate about whether the rush of new policy initiatives has been business-as-usual or a more radical departure. This uncertainty reflects the conflicts within the British New Right between the neo-liberal eco-optimists (such as Beckerman 1995 and North 1995), and the apparently more traditional communitarian position of Gray (1993) and others: the former have argued that the market is the best allocator of scarce resources such as environmental goods, the latter that there is a strong case for intervention consistent with conservative principles to protect public goods such as environmental assets and community values, and to promote risk-averse behaviour.

Partly, too, these contradictions stem from the difficulties the UK government has had in continually needing to respond to the environmental agenda set by Brussels, while maintaining its position that, on the principle of subsidiarity, the details of environmental policy are best left to the level of national government.

A further area of contradiction is in the nature of British land-use planning itself, and in particular the ability of the system as defined and constrained in legislation to take on issues broader than just the use of land. To some extent, this reflects the tension between local government and the confining role of central government, keen not just to limit the possible powers of an alternative tier of the state, but also to avoid overlapping bureaucratic controls. It also reflects the reluctance of the profession as an association to defend the role of planning, and notably shows the new role of environmental pressure groups in the late 1980s and early 1990s as defenders, promoters and sources of best-practice advice for local planning authorities.

To illustrate these points, I shall in this chapter focus on two areas of new activity for land-use planning in the environmental field: (a) the relationship between planning and pollution control (such as the new environmental regulatory bodies, and the acquisition of new planning responsibilities in the fields of waste planning and air quality), and (b) the environmental appraisal of development plans.

I shall conclude by arguing that the changes in the 1990s, despite these contradictions, do represent a very real shift in the focus of land-use planning activity consistent with both conservative and radical value-systems, and one which I therefore expect to persist into the next century.

Environmentalism and the British New Right

Before examining how land-use planning has responded to any new role in delivering more environmentally aware decisions, it is important to recognise the broader political and ideological debates about the adoption by central government of the environmental agenda.

The very name of the government's first environmental strategy, *This Common Inheritance* (Her Majesty's Government (HMG) 1990), and the justification given in the document itself, were classic statements of a certain sort of conservatism based on stewardship, consistent with the traditional paternalistic conservatism of the shires and landed estates, and harking back to nineteenth-century Toryism, which might seem at odds with the neo-liberal, market-oriented view of conservatism. However, in the attempt to get all wings of the governing party to sign up to the document, which bears the signatures of all the secretaries of state for the chief government departments – a considerable feat for the then Secretary of State for the Environment, Chris Patten – the strategy also gives expression to the neo-liberal view, particularly in the emphasis on voluntarism and market-based instruments for delivering environmental policy. British environmental policy itself in the 1990s can therefore be seen as a struggle between these two strands in conservative philosophy.

These dilemmas are nicely expressed by John Gray in his essay allying green thinking and conservatism (Gray 1993). In it, he argues that the two positions naturally converge, in that traditional Burkean ideas of a social contract envisage such a contract as entailing responsibilities to the past and to future generations, as do green advocates of multi-generational sustainability. The traditional conservatives, he argues, are also sceptical of the concepts of progress and innovation, and he supports the precautionary principle (one of the guiding principles of *This Common Inheritance* and its successor, *Sustainable Development: the UK Strategy Plan* (HMG 1994b)). Gray advocates a conservatism which recognises a need for strategic planning to limit the market in order to promote 'the common life' (Gray 1995: 137) and to deliver such ecological public goods as biodiversity.

What we see, therefore, in the development of environmental policy in the 1990s, is a response to fundamental conflicts between the ideologies and aims of conservatism – between the need for deregulation, lifting the burden off small businesses, promoting a culture of enterprise, individual initiative, competition and privatisation, consistent with the neo-liberal wing of the party, and a traditional view that we should conserve past investment in infrastructure such as rural communities, and even in town centres and cities, through restraining the forces of commercialism.

How far these different views have been given expression in Cabinet debates, as opposed to the think tanks of the New Right, is not clear; but it helps to explain the two strands in conservative environmental policy in the 1990s: the emphasis on voluntarism, environmental education and self-awareness which is such a feature of *This Common Inheritance* and the *Sustainable Development* strategy; and the continuing flow of primary legislation and secondary instruments which, especially in the fields of waste and pollution, impose real constraints on the operations of businesses. These contradictory philosophies account for some of the variety of measures which were being tried out in the 1990s, and provided relatively fertile ground for the range of local initiatives in the field of land-use planning, discussed in depth later.

The influence of Europe

A different source of stimulus for the new approach to environmental policy was the need to co-ordinate a response to the steady flow of environmental legislation emanating from the European Communities, the need to have in place some framework to respond to, and if necessary pre-empt, these requirements and other international obligations, and the perhaps belated realisation that the passage of the Single European Act in 1986, and the Maastricht Treaty on European Union in 1992 (Commission of the European Communities (CEC) 1992) exposed the UK to the possibilities of yet more environmental legislation under the wider provisions of qualified majority voting in the new regime.

Europe has adopted a common position on the environment for good reasons – the fact that environmental impacts do not observe boundaries, the need for a level playing field for the free movement of people, goods and services, and the need for the EC itself to respond to international commitments such as the Rio Summit. Despite there being no specific mention of an environmental remit for the Communities in the original Treaty of Rome, the policy has evolved and expanded (Haigh 1992). Following the Stockholm Summit of 1972, the EC prepared successive Environmental Action Plans, with the Fourth running from 1987 to 1992, and the Fifth from 1993 to 2000.

Figure 2.1 shows the huge increase in legislation – directives, regulations and opinions – which emerged from Brussels under these plans over the period 1985–95: over time, the approach changed from measures regulating certain processes and emissions, such as large combustion plants, to much broader-ranging procedural directives such as those on the *Assessment of the Effects of Certain Public and Private Projects on the Environment* (CEC 1985), and *Freedom of Access to Information on the Environment* (CEC 1990). The former of these had to be implemented in 1988, and the latter in 1992.

Part of the UK government's case in the 1970s against the environmental assessment legislation had been that it was already in place in the British town and country planning system (Wood 1988); but as the flow of European legislation in the environmental field continued in the 1980s, they argued that it was inconsistently implemented across the different member states, and poorly monitored and enforced. The UK was therefore, for consistency with its arguments against extending these provisions, obliged to show that it was at least not a 'lag state' as far as implementation was concerned.

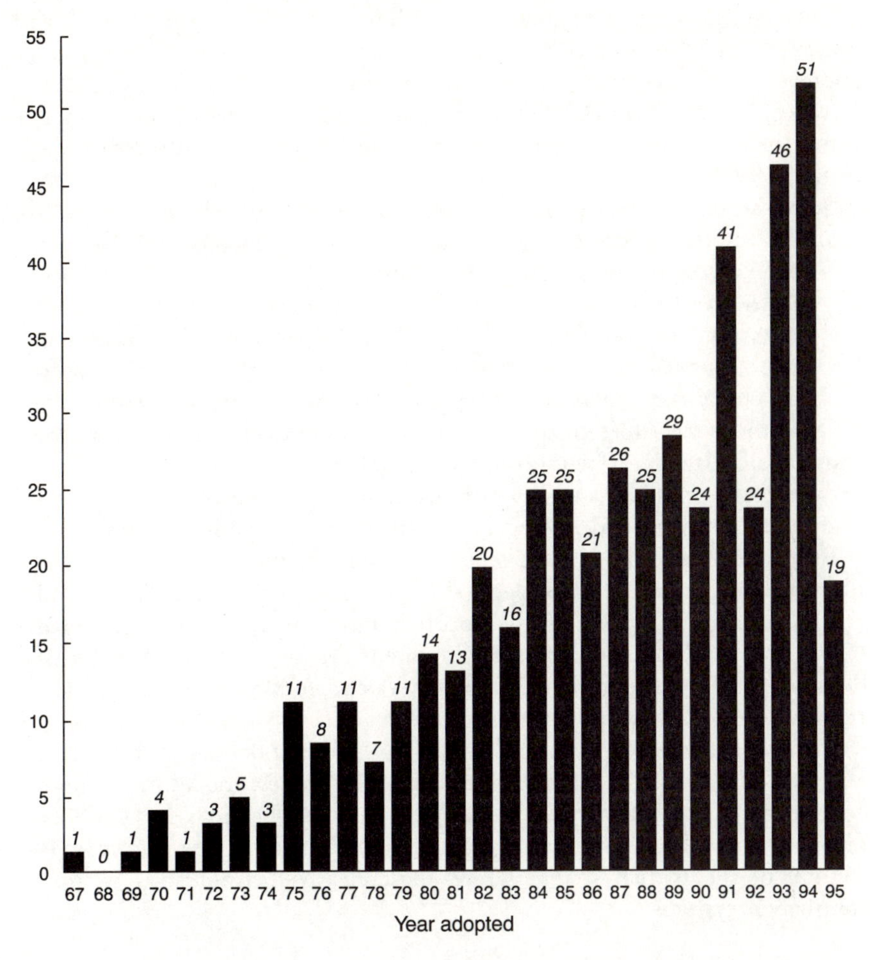

Figure 2.1 Number of items of EC environmental legislation adopted each year
Source: Haigh (annual)

Changes in policy styles

In his book reviewing the character of environmental policy in the 1990s, Tim Gray argues that it supports two possible interpretations. The first is that 'the period has witnessed a fundamental sea-change in UK environmental policy-making'; the second that 'any change has been merely "rhetorical"' (Gray 1995: 1–2). In support of the latter, it was argued at the time that *This Common Inheritance* merely expressed comfortable words and good intentions, with no resources, no policy shifts and no clear targets necessary to implement it (O'Riordan 1990). In support of the former, it has been argued that the traditionally discretionary, pragmatic approach of government to environmental policy (as described, for instance, by McCormick 1991) has begun to be replaced by a more comprehensive, integrated approach more akin to continental policy styles (Burke 1995).

Some argue that it is the European dimension which has 'radically changed the evolution of British policy-making' (Haigh and Lanigan 1995: 29). Despite some apparent continuities in the style of UK policy-making, they contend that the EC has fundamentally influenced both the process and the outcomes of UK policy: in particular, 'centralism in UK environmental enforcement is directly linked to the increasing decisions on policy content at the European level' (Haigh and Lanigan 1995: 30). At the same time, there has had to be more detailing of thresholds and emission standards at the European level, partly to ensure consistency between lead and lag states, which has removed some of the scope for traditional lobbying and behind-doors negotiation in pollution control.

Even if the late 1990s have seen a faltering in the rush of legislation illustrated in figure 2.1 (as suggested by, for example, Kramer 1995, and the EC itself in its reports on the implementation of the Fifth Action Plan (CEC 1994 and the European Environment Agency (EEA) 1995)), this does not alter the conclusion that past developments have forced substantial changes to British policy-making.

The consequence of this is that, for all its expressed belief in subsidiarity and flexibility, the UK has had to adapt its policy modes to a more continental style. This of course has had clear implications for land-use planning as well as environmental regulation, because it has exposed the features of the British system which most clearly distinguished it from continental systems.

A receptive system?

Despite the ideological and political tensions over developing environmental policy, and hence the scope for land-use planning, it can also be argued that the seeds of change had already been sown in the 1980s. Major legislative commitments were in place before the two key events often cited as heralding a new phase – Mrs Thatcher's speech to the Royal Society in 1988, and the launch of the UK government's first Environmental Strategy in 1990 (HMG 1990). The Environmental Protection Bill, for instance, which brought in substantial changes to the regulation and disposal of waste, was drafted in response to the recognised inadequacies of the Control of Pollution Act of 1974.

At the same time, international negotiations over climate change were not only causing a reappraisal of the relationship between land-use planning and transport (DoE 1992a; DoE/DoT 1993c), leading to a revision of PPG 13 (DoE/DoT 1994), but they also introduced the idea of explicit target-setting in the environmental policy area. The 1990 Environmental Strategy set targets, initially in such areas as waste recycling and renewable energy, which have been continued in the mid-1990s with targets in new areas such as cycling, local air quality, and the proportion of homes to be accommodated in existing urban areas.

Of course, the setting of targets in many areas has itself reflected a preoccupation with performance indicators, giving expression to the perceived need to make the public sector more accountable and more in line with what was held to be good practice in the private sector. But these targets have resonated with Local Agenda 21 work on local environmental indicators (such as Strathclyde 1995 and Fife 1995), and have had considerable implications for land-use planning through the integration of conventional monitoring of development plans and development control with targets set in the plan (such as Bedfordshire 1995).

These changes – an increase in legislation governing waste, the use of targets for atmospheric emissions, and an accepted reappraisal of the land-use–transport relationship – were clearly flagged at the beginning of the decade as implying substantial shifts in the traditional focus of land-use planning. The signs of change were also evident in the land-use planning system itself, with the Nuffield Report on the town and country planning system in 1986 (Nuffield 1986) pointing up two particular areas for change: the need for clear government guidance in terms of policy (as opposed to the issuing of circulars which interpreted legislation), and the need for firm guidance from the development plan.

Planning policy guidance and the plan-led system

The model which Nuffield cited for government guidance was the Scottish system of national planning guidelines, and although many argued that the policy community for planning in Scotland was very different from that in England and Wales, with more shared education and without the central–local antagonisms, the suggestion was taken up by Whitehall. The system of Planning Policy Guidance Notes (PPGs) was introduced in 1988. Whether or not this was seen at the time as an opportunity for increased centralisation and control of local decisions, it had the effect of providing a notable vehicle for the government in the 1990s to push through its interpretation of sustainable development in planning without any change in the legislative framework.

By the end of the 1980s, the revival of the plan-led system (discussed in chapter 5), and the opportunity for self-certification in structure plans, could be seen as encouraging innovative plan-making which could take on a broader environmental perspective.

What we see therefore is a combination of apparently contradictory circumstances which produced a system which seemed wide open to the adoption of new forms of plan-making: at the same time as local authorities' room for initiative seemed more proscribed, new possibilities were opened up for making the scope of planning more synoptic and innovative.

How far has the planning system therefore been able to take on the environmental agenda of the 1990s? And, as with the claimed shift in UK environmental policy generally, is this anything more than comfortable words?

A new environmental stance for planning?

There is a broad spectrum of views amongst practitioners and observers of the planning system on how far, and indeed whether, British planning has taken on a new environmental stance, ranging from the enthusiastic advocates of a new role for planning to those who see it as wishful thinking or self-delusion in the context of unchanged political and economic reality.

At one end is the view that planning has taken up the new area of sustainability as a consistent extension of its traditional concerns. On this view, planning in the whole of the post-war period up to the late 1980s entertained a very limited notion of the environment, confining itself primarily to concerns with 'amenity', especially visual amenity. It expressed a conception of the public interest unrelated to any assessment of the physical or natural environment, except for some attention to a small range of pollutants such as noise and dust (Newby 1990). Nevertheless, it possessed other characteristics which fitted closely with the new sustainability agenda.

The proponents of this extended environmental role for planning in the 1990s (e.g. Hebbert 1992; Millichap 1993; McLaren and Bosworth 1994) advocated a view of sustainability which they argued was fully compatible with some of the precepts of land-use planning – sustainability broadly defined as consisting of the four principles of intergenerational equity, community participation, long-term horizons, and a broader definition of development than one based merely on economic wealth conventionally defined as per capita GDP. They argued that planning had the powers and remit necessary for pursuing these objectives: it had the power, certainly at the level of structure planning, to look ahead, even if only for the next fifteen years rather than for succeeding generations; it had some experience of resource planning and management in its responsibilities for minerals planning, and indeed for land itself as a scarce resource; it was familiar with the need to reconcile competing social as well as economic objectives in its interpretation of development, however constrained by central government; and, finally, planning had, from the 1960s, been associated, however fitfully and uncertainly, with public participation in decision-making on both plans and projects, and with community development as an end in itself.

This model of adaptation is supported by Rowan-Robinson *et al.*, who contend that planning powers and tools have generally been able to ensure that land-use objectives 'are compatible with sustainable development' (Rowan Robinson *et al.* 1995: 283). According to this view, this shift has made a difference not only to the process but also to the outcomes of planning.

Others argue that, despite these good efforts, planning's success in terms of outcomes is inevitably limited by both internal and external factors. Michael Welbank, then president of the Royal Town Planning Institute (RTPI), expressed the fear that central government had 'dumped' environmental concerns on the land-use planning system without giving it the necessary powers of implementation (Welbank 1992 and 1993). Similarly, Healey and Shaw (1993) contend that these limits are imposed by planning's discretionary nature, by its inability to deal firmly with public as opposed to private development (Healey *et al.* 1988), and by a continuing inability to see policy intentions through to well-audited implementation. They also point out the constraints of key responsibilities for environmental policy being fragmented amongst different non-elected quangos, such as for agriculture (ADAS) and water (at that time the National Rivers Authority – NRA), the lack of technical certainty in areas such as the relationship between energy consumption and urban form, and the fact that where, in a highly centralised system, innovations at the local level had been initiated (such as in Kirklees and Sutton), these had come from outside the planning function. Brindley *et al.* ascribe the shift from the trend planning of the 1980s to the more responsive style of the 1990s partly to the new environmental agenda, but this new style still has to respond to market forces (Brindley *et al.* 1996).

A more critical extension of this argument is that the planning system and profession have not changed in any fundamental way, and hardly in any superficial way, in that the key characteristics of the system, both in terms of the framework within which it operates and the outcomes of that policy, are determined by the state acting in accordance with the promptings of highly mobile capital (Marshall 1994). The system's continued emphasis on land use narrowly defined, its inclination to allow socio-economic arguments in favour of a particular form of capitalist or state development to override environmental objectives, and its inability to maintain absolute protection for any key biodiversity site or habitat, in the face of other material considerations such as the national economic interest, are all cited as evidence of the narrow scope for real change. In this view, neither the processes nor the outcomes of planning are any more likely to favour environmental concerns; rather, the scenario is business-as-usual, in the same way that characterises Britain's environmental policy.

Tim Marshall, for instance, warns that, while the current challenge to central government's narrow land-use conception of planning could be seen as similar to that of the community planning movement of the 1970s, it faces similar difficulties. True environmental planning really requires 'social redistribution and social (re)empowerment' to gain democratic support, and integration with fiscal and other regulatory measures to have the force to change corporate behaviour (Marshall 1994: 28).

This chapter therefore aims to examine the currency of these different perspectives as we approach the end of the decade. The first area I shall use as an example to illustrate the breadth and variety of locally generated experience is that of the relations between planning and pollution.

Planning and pollution

On all but the first of the dimensions of planning and the new environmentalism discussed above, there should have been no change in planning's traditional inability to deal with pollution broadly defined. Miller and Wood had described the awkward relations between land-use planning and the pollution control agencies in the 1970s and early 1980s (Miller and Wood 1983). Although planning had been given some responsibilities to consider potential pollution, for instance in the areas of potentially hazardous major developments (DoE 1984a), and later potentially hazardous substances (DoE 1992c), confusion and mistrust remained. By the beginning of the 1990s, new quangos in the form of Her Majesty's Inspectorate of Pollution (HMIP) and NRA had been created, and given new powers in waste and integrated pollution control under the fairly innovative Environmental Protection Act (EPA) of 1990, which appeared to take pollution concerns further from land-use planning's remit.

It was clear that, far from expecting planning to take on a broader concern with pollution, the government wished there to be a clear demarcation of responsibilities, with no overlap, for the sake of deregulation, minimising bureaucracy, making clear lines of responsibility and transparency for the business community, and possibly to prepare agencies for possible privatisation. Nevertheless, they recognised that the 1990 Environmental Strategy required attention to be paid to the scope for integration between different regulatory regimes. The EPA 1990, with its many new provisions relating to pollution and waste, and procedures for preventing or minimising releases to air, water and land, necessitated an examination of the potential for overlap between the regulators of these media. Moreover, there were other changes which challenged any neat separation, such as the implementation of the European Environmental Assessment Directive through the Town and Country Planning Regulations of 1988, which specifically required LPAs to take into account, *inter alia*, the likely significant effects of major projects on 'soil, water, and air'.

Research on planning and pollution

Faced with this new complexity, and with evidence of growing public concern over the relationship of development and pollution, as well as between the different agencies, the Department of the Environment commissioned research into the issue (DoE 1992c). The report of the research cited five grounds (p. 6, para. 2.2) for these concerns:

1 growing public awareness of pollution issues, expressing itself in concerns addressed to elected local authorities 'which had the advantages of ready accessibility and responsiveness to democratic pressures, which it is perceived the pollution control authorities do not have';
2 perceptions by the public and by local authorities of the pollution control agencies as having been historically over-considerate of industries' needs;
3 perceptions that the pollution control agencies interpret risk too narrowly, ignoring, for instance, 'the effects of cumulative pollution from incremental development, the risk of pollution from unintended releases, the impact of perceptions of risk on the local community and on investor confidence, and the risks of non-compliance with pollution control measures';
4 resultant pressure on planning authorities to give pollution risks as reasons for refusal of planning permission, with consequent scope for challenge, uncertainty and delay; and
5 the implementation of the EIA Directive, providing more information than before on the impacts of projects, and therefore raising awareness.

The research concluded that 'all of these factors have contributed to the increased readiness of local authorities to incorporate policies to prevent pollution in development plans, to use pollution reasons as grounds for refusing planning consent, and to seek to control pollution from development through conditions and planning agreements' (DoE 1992c: 6).

Development plan policies

The research found that an additional reason for the increase of pollution policies in development plans was a result of local authorities' conducting or commissioning environmental audits of their areas, and looking to the development plan as a ready vehicle for setting out policies and priorities for action. This move towards state-of-the-environment reporting had been prompted by the Friends of the Earth *Environmental Charter for Local*

Government (McLaren and Adams 1989): by 1992, a good proportion of local authorities had conducted or were conducting such audits (Wilson and Raemaekers 1992), and the figure has increased considerably with the work being done as part of the Local Agenda 21 process in other authorities (Morris 1995).

This finding therefore suggests a revision of the view that no substantive changes were being made to the form of policies or their implementation at the local level. At the same time, of course, the UK had moved towards a more explicitly plan-led system, as discussed earlier, which enhanced the status of policies, and perhaps meant that planning authorities were more prepared to make explicit response to the concerns of their residents over environmental quality.

Such policies were broadly of two types. There were the traditional locational polices, which nevertheless were extending the range of their definitions of sensitive environments: such policies included (DoE 1992c: 23, para. 3.6.3):

1 those defining sensitive areas (such as aquifers, which had not formerly been a concern of LPAs) where potentially polluting development should not be permitted;
2 those defining sensitive uses which should not be permitted in the vicinity of existing potential sources of pollution – again, such uses were defined more broadly to include not just housing but also schools and in some cases offices;
3 those allocating land as suitable for potentially polluting development.

There were also general policies opposing development which might give rise to any or unacceptable levels of pollution: and in these cases too the research found that the type of pollution being addressed was altering, embracing not just the traditional focus of local authority concern with noise, fumes and dust, but also for the first time both more specific pollutants such as leachate or migrating gas, and more general pollutants such as other emissions to air (DoE 1992c).

In this area, it would be very hard to maintain that local planning authorities were feeling over-constrained in their ability to take on the much broader conception of environment which the new agenda required. The value of the research lies in showing from detailed examination of the development plan documents and development control decisions just how far apart local practice had become from the expectations of the centre. The research concluded that elected local authorities, democratically accountable to the electorate, had for a variety of reasons taken on the new environmental agenda and absorbed it within the well-accepted planning framework. As much as showing the scope for local change, it also revealed the very different and broader perceptions of environmental risk which local authorities had, compared with the pollution control authorities (Wilson 1996). These findings were confirmed in a separate study (undertaken earlier) of the relationship between LPAs and another agency, the Health and Safety Executive, and the then HMIP, over consultation on chemical installations, in which Miller had similarly concluded that planning authorities' judgements about the pollution risks associated with such developments were very often neither based on nor in line with the advice of the pollution control agency (Miller 1994).

PPG 23

The Department of the Environment research was intended to inform the Department in drafting a PPG on planning, pollution and waste. This drafting proved problematic, and the final version, PPG 23 (DoE 1994b), came out only in 1994, to claims that it failed to resolve with any clarity the relationship between planning and the other control agencies (Weatherhead 1994; Weston and Hudson 1995). PPG 23 reaffirmed that 'the planning and pollution control systems are separate but complementary' (DoE 1994b: para. 1.2), but, while it reiterated the conventional departmental view that 'the planning system should not be operated to duplicate controls which are the statutory responsibility of other bodies' (para. 1.3), it also had to accept that 'the dividing line between planning and pollution is not always clear cut' (para. 1.34).

The solution to this problem proposed in PPG 23 is to provide information to all parties on the scope of their respective responsibilities, and to recommend extensive consultation on applications for potentially polluting development, and on the formulation of policies in development plans. The guidance on interpretations of risk, in particular, however, remains ambiguous (Weston and Hudson 1995; Wilson 1996), and there is little evidence that the PPG has deterred planning authorities from including explicit pollution-related policies in their development plans. Miller sees this as only to be expected: 'similar advice, admonition and exhortation in the past have signally failed to curb planning authorities' readiness to use their statutory powers to secure what they consider to be the appropriate level of control over pollution sources which other agencies are unable or unwilling to enforce' (Miller 1994: 128).

How far then has the PPG been effective in persuading planning authorities to rely on these other agencies? A brief survey of some development plans prepared since the PPG shows that many continue to include policies making explicit reference to pollution. While some rather unspecifically extend the traditional list of pollutants (such as 'In considering development proposals the council will not permit schemes unless it is satisfied that there will be no conflict arising from vibration, smell, smoke, fumes or other forms of pollution' (Vale of White Horse 1995: 217)), many other LPAs include more explicit policies, such as Dinefwr Borough Council's draft deposit local plan policy DLPEN2:

> It is the policy of Dinefwr Borough Council to protect those basic, natural elements that interact and contribute to the quality of the landscape and environment of the local plan area. Any development which will be detrimental to the overall amenity value, nature conservation interest, water quality, soil quality, air quality, hydrological and geological regimes will not be permitted.
>
> (Dinefwr 1995: 24)

Planning and the pollution control agencies

One of the reasons for the reluctance of LPAs to rely on the HMIP and the NRA for pollution advice was their perception that the results of the consultation were often too slow, vague or standardised, and that the control agency would not be prepared in any case to provide support if it came to appeal. This applied particularly to HMIP, where the 1992 research found that 'the history of its predecessors, its apparent remoteness from local government, its recent establishment', and its more immediate concern with putting its own guidance in place, militated 'against good collaboration between planners and the Inspectorate' (DoE 1992c: 20).

This contrasted with the NRA, with whom planning authorities had in some areas established good relationships, as with its predecessors: '[a]longside this it has always been in NRA's interest to use the planning system to achieve certain of its objectives and NRA regions tend to take a very pro-active stance on liaison with planning' (DoE 1992c: 20). An example of this pro-active approach throughout this period was the Thames Region NRA (NRATR), which pioneered some highly innovatory developments which, while aiming always to work with the planning system, in some senses can be seen as a potential threat to its sphere of influence (a feature of the context within which planning operates to which Healey and Shaw (1993), Marshall (1994) and Slater *et al.* (1994) draw attention).

NRA interaction with the planning system

Two aspects of NRATR's work merit attention here: the development of what were originally termed 'model policies' for the water environment, which the NRA promulgated in order to protect 'its' interests (that is, those of the water environment) in land-use planning decisions; and the preparation of Catchment Management Plans. The NRA was keen to do this as it recognised that its very broad range of functions (landscape and nature conservation, flood protection, water resources and quality, fishing, recreation and navigation) were strongly related to many land-use issues, and the terms of its licensing and consent powers did not cover all developments which might impact on, for instance, ground water resources. While it was accepted that much pollution to ground water supplies arose from agricultural developments outside the remit of planning, the NRA also saw the opportunity to provide some form of strategic input to other locational decisions in order to protect, for instance, boreholes and flood plains.

The idea for draft model policies originated in the Thames Region in response to the opportunities presented to the NRA to take a more proactive role in the light of the requirement that all London Boroughs prepare Unitary Development Plans (UDPs) (Gardiner 1994). They were tested in the Thames and other Regions and finally issued as guidance in 1994 (NRA 1994).

This guidance sets out the NRA's interests in relation to strategic plan-making (structure plans or UDPs Part I), and local plans or UDPs Part II. The Dinefwr hope was that 'the objectives will be substantially replicated in each local authority's land use plans as they are appropriate', while recognising they might need some modification. The guidance promised that 'The NRA will subsequently assist with the implementation of policies where appropriate, either through the consultation process or by the use of its statutory powers and in the execution of its duties. Conversely, the NRA may formally object to development plans which conflict with the stated objectives' (NRA 1994: 2).

NRATR in London had undertaken a very revealing study of the success of this proactive approach in the uptake of the original draft policies in development plans in London: figure 2.2 shows that, by 1993, most boroughs had included over 50 per cent of the policies, and three – Brent, Merton and Tower Hamlets – had achieved a 100 per cent rate. Subsequent unpublished research shows that the take-up of the guidance policies in the Thames region is similarly comprehensive (Reynolds n.d.).

In this area, therefore, far from the planning system being elbowed out as a locus for action on environmental concerns by the quangos, it can be argued that it was utilised in what seems to be a highly effective, proactive way by a dynamic region.

HMIP interaction with the planning system

Compared with the NRA, HMIP had been much slower in seeing the opportunities for a proactive approach with respect to land-use decisions, and only belatedly, in 1995, issued guidance notes on liaison with planning authorities (HMIP 1995). The research commissioned by the Department of the Environment on the relations between planning and pollution found that HMIP was much less prepared to see the potential of collaboration with the

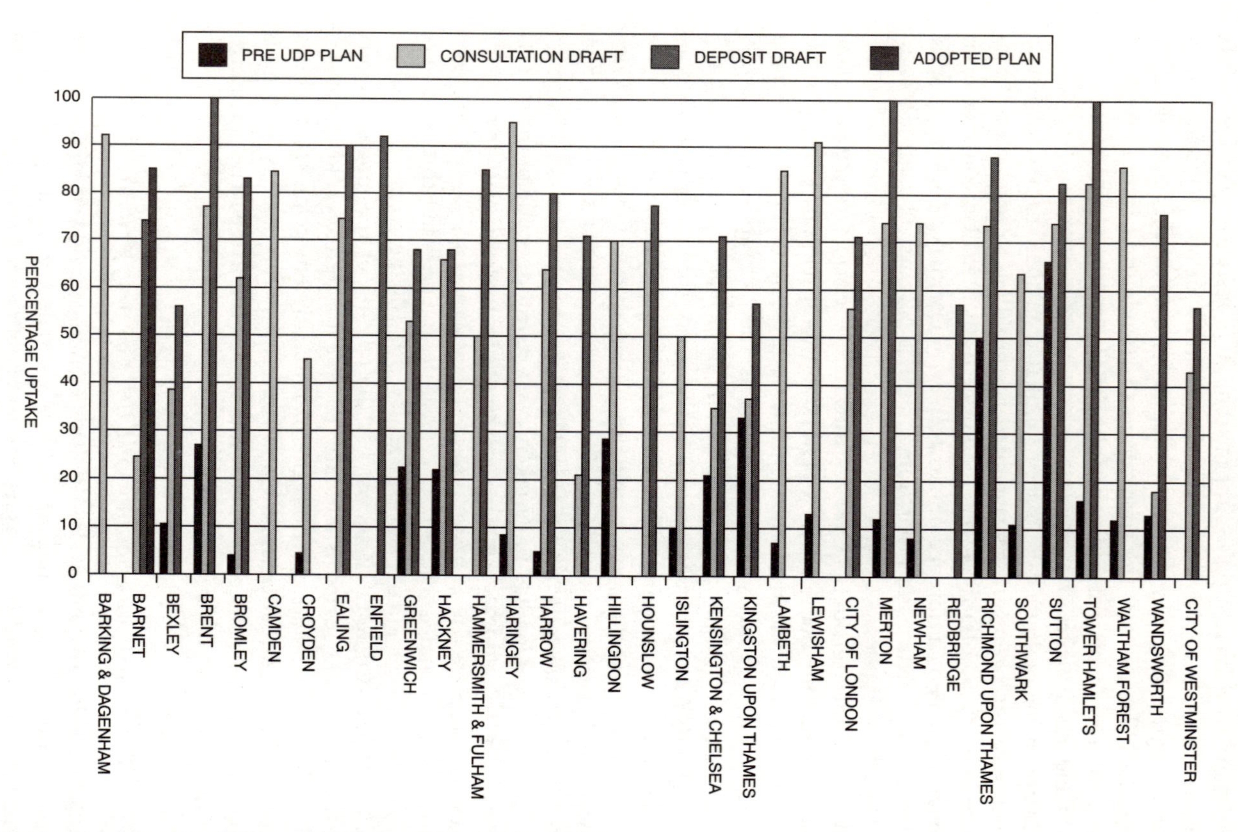

Figure 2.2 Uptake of model land-use policies (as at Spring 1993)
Source: Gardiner 1994

planning system, and maintained a firmly technocratic view of issues of pollution (DoE 1992c). This was to some extent explicable, as discharge consents under Integrated Pollution Control (IPC) were often negotiated by developers after any planning consents – an issue which still gives rise to concern over the quality of assessments of the impacts of some major projects (DoE 1996a) – but it also reflected the very different regulatory ethos of the organisation (Carter and Lowe 1995).

Nevertheless, signs of change were evident in the mid-1990s. In response to work by Kent County Council in modelling air quality for the North Thames coast (Kent CC 1994), HMIP had been prompted to adopt a more strategic view for the whole East Thames corridor in assessing the possible cumulative impacts of new industrial and waste processes (HMIP 1995). In the 1995 guidance, HMIP committed itself to further regional environmental assessments – a point discussed further below in looking at the production of local authority Air Quality Management Plans and the environmental appraisal of development plans.

Catchment Management Plans

A second aspect of the NRA's innovation which could be seen as more directly in competition with the scope of planning was the development of Catchment Management Plans (CMPs): by the time of its amalgamation into the Environment Agency in April 1996, NRATR had produced eight of these, covering half the region. While each one was launched with an extensive programme of consultation with other agencies, interests and user groups, the scope of the plans presents a very real challenge to planning authorities, as they have no statutory status, and yet were intended to govern NRATR's actions including its response to planning consultations (Slater *et al.* 1994).

In Thames Region, in particular, an attempt was made to draw these separate CMPs together with a review of development plans in the region in their *Thames 21* document (NRATR 1995). *Thames 21* had three aims: to help to articulate water interests at a strategic level, such as in Regional Planning Guidance; to provide a convenient summary of policies for the development plan system; and to provide a regional context for CMPs with an indication of development pressures. *Thames 21*'s map of the areas of water-supply concern in relation to their view of development pressures for the northern sector of the region is shown in figure 2.3.

This shift towards planning at the level of natural resources might be welcomed by many environmentalists as a move towards real resource management planning. Conversely, it can be argued that, unlike local planning authorities, the NRA lacked any political accountability, could call on considerable financial resources (because it had functions of development as well in the areas of flood control and river management), had

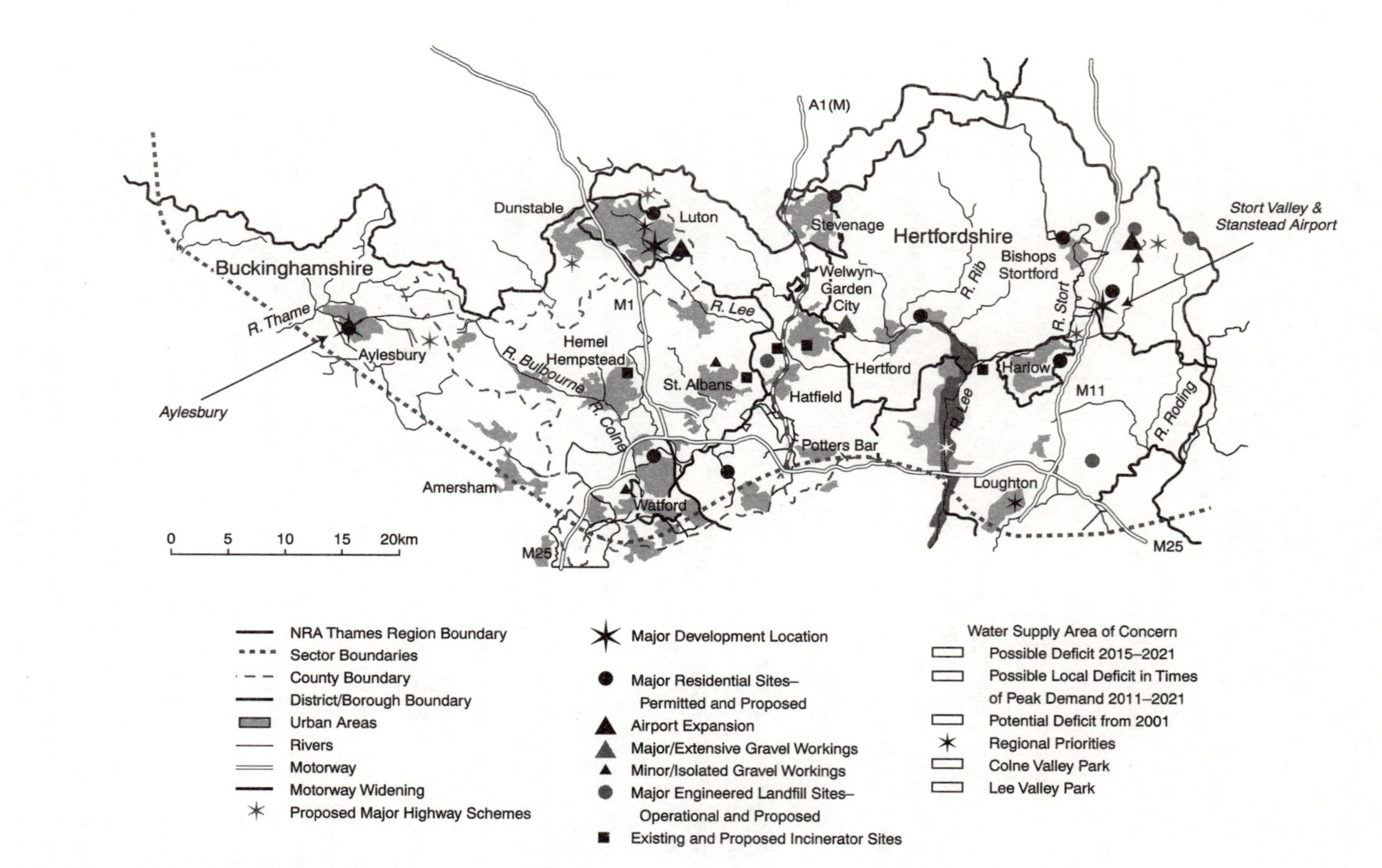

Figure 2.3 Pressures and areas of water-supply concern
Source: National Rivers Authority Thames Region 1995

the ability to operate at a more strategic level, and had powers of direction, and consequently posed a real threat to the integrity of planning.

Local Environment Agency Plans

As many commentators have noted (e.g. EDS 1996), the NRA seemed to win out in the formation of the new Environment Agency in 1996, in terms of key personnel and the weight of resources which it brought. In the early, exploratory months of the new regime, it seems that integration between the former agencies may be hard to achieve, but that the NRA culture will dominate. This might favour the maintenance of good relationships with planning authorities; but the proposals to extend the CMPs to cover all Agency functions in new non-statutory Local Environment Agency Plans (LEAPs) may offer a very real challenge to any new-found environmental scope in land-use planning.

LEAPs are intended to draw together the different responsibilities of the Agency through setting out a common vision for an area: this involves setting priorities for the Agency through reviewing the local environment, identifying the key issues for that area, and establishing an integrated plan of action of managing the area for a five-year period.

An early example illustrating the process and content of these plans is the *Consultation Draft of the Local Environment Agency Plan for the Thames (Buscot to Eynsham)* (EA 1996). This LEAP identifies as key issues substantive land-use topics such as urban development, mineral extraction and road construction. With the possible overlap of the scope of statutory development plans, and possible duplication of the public consultation process, there appears to be a real chance that environmental planning in the late 1990s will be more fragmented.

In Scotland, in particular, with the loss of the strategic tier of planning authority with the abolition of most of the Regional Authorities, the development of equivalent catchment-based plans by the Scottish Environmental Protection Agency may well provide a direct challenge to the integration of environmental concerns at the level of strategic land-use planning in fragile 'joint working' arrangements (Lloyd 1996), particularly as the responsibilities for water and sewerage have been removed altogether from local authorities.

Planning and air quality

Other developments in local authorities' responsibilities, however, suggest an alternative interpretation. A further example of new responsibilities for the planning system which I want to consider is that of the implementation of the national air quality strategy. The prompts for this strategy have come from a number of sources: the international commitment under the

Climate Change Convention to limiting CO_2 emissions, the threat to environmental assets such as biodiversity from emissions to air, and the growing concern with the environmental health consequences of traffic pollution, and in particular the risks to children of increased bronchial illnesses. Whether these are directly caused by pollution from motorised traffic, or whether they are triggered by the incidence of pollutants such as nitrogen oxides and volatile organic compounds, they nevertheless pose risks to the nation's health, and imply further costs to the budgets of the National Health Service, which are already under pressure. The government was accordingly convinced of the need for precautionary action – despite what it claimed was the evidence from monitoring of the improvements in air quality (DoE 1992a).

The government's strategy for air quality management (DoE/SO 1996c) proposes three principal approaches: setting a framework of national air quality standards and targets; developing new systems for local air quality management on an area-wide basis; and implementing stronger controls on vehicle emissions.

These involve awarding substantial new responsibilities to local authorities in reviewing air quality in their areas, and developing Air Quality Management Plans for remedying the position in problem areas. They also imply integrating planning policies and transport strategies, in line with PPG 13, to allow for a reduced need to travel and to encourage a shift to less polluting modes of travel. These air quality reviews and plans are not likely to be prepared under the umbrella of the development plan process (DoE/SO 1996c), but will co-exist, perhaps rather uneasily, with them.

To some extent, far from being seen as a radical new departure, these responsibilities will rather simply confirm the approach which many authorities are already taking. Despite the admonitions of PPG 23, evidence of planning authorities' concerns with air quality can be readily found in the recent crop of development plans for both metropolitan and other areas. Almost all the plans so far mentioned in this chapter include a policy on air quality, such as Devon County Council's draft Structure Plan review Policy EN28: 'Development that would give rise to an unacceptable deterioration in air quality should not be permitted' (Devon CC 1995: 47).

At the same time, there are other implications for the land-use planning system. In addition to the need for periodic review of assessments under the AQMAs, such as at the time of revision of land-use development plans for the area, the draft strategy confirms that local authorities should 'appraise development plans and transport plans against these detailed assessments of air quality' (DoE/SO 1996c: para. 7.30). As discussed below, one of the difficult areas for the environmental appraisal of development plans has been the level of knowledge of baseline conditions against which to appraise the policies and proposals of the plan. The availability of information from detailed air quality assessments will offer scope for better integration of other local authority functions and expertise in the appraisal process, and possibly encourage a radical approach towards integrated environmental management across all the functions.

A number of urban and metropolitan local authorities had already moved towards this integrated approach in their innovative work on local energy strategies (such as South Glamorgan and Cardiff (South Glamorgan 1994) and Newcastle (1992)), prompted by resource conservation concerns as well as by air quality issues.

It also seems inevitable that local authorities will need to co-ordinate their locally based Air Quality Management Plans with the Environment Agency's work in preparing the air quality element of LEAPs, and that the Agency will need to work with regional planning conferences at the strategic level, particularly as air pollution raises major trans-boundary issues. The cautionary note of PPG 23, with its emphasis on discrete functions for different pollution control regimes and its advocacy of the virtues of mere consultation between different regimes, seems at odds with the more integrated approach of the Air Quality Strategy.

Planning and waste

In other areas of planning's remit for protecting and managing environmental resources, yet other principles are being established. PPG 23 has been redrafted to take account of the new Waste Strategy (DoE 1995) and of the inclusion in the Environment Agency of the waste regulation function, formerly under local authority control. Two new principles are being introduced: the proximity principle and regional self-sufficiency (DoE 1996b).

These are, as yet, new principles, but they illustrate the extent to which the land-use planning system, at the same time as displaying initiatives in pollution policies at the local level, is also being expected by the developing national agenda to adopt new approaches. The argument that there will be a forcible diminution of local planning authority interest in the environmental agenda in the second half of the 1990s seems therefore unconvincing.

On the contrary, it seems reasonable to expect that land-use planning in the late 1990s will see a change not just in its processes but also in the outcomes on the ground, particularly in urban areas. It is the contention of this chapter that the scope for locally generated innovation in adapting planning to the new environmental agenda has been created by the ambivalence of the New Right towards this agenda. The centralist, budget-capping tendencies, and reluctance to use supporting economic instruments, have provided the prompt for local action to show its independence of the centre, at the same time as providing the indicative policy guidance through waste and air quality strategies, and PPGs, to support local initiatives.

Environmental appraisal of development plans

The second area in which local autonomy can be tested is the implementation by LPAs of the expectation (it is arguable whether it is a formal requirement of the plan preparation process) in PPG 12 (DoE 1992e) that LPAs appraise their policies for their environmental implications.

This part of PPG 12 was intended to satisfy both the commitment in *This Common Inheritance* to take fuller account of the environment in governmental decision-making, and to fulfil the requirement in the Planning and Compensation Act 1991 that authorities have regard to environmental considerations in preparing their structure and UDP (Unitary Development Plan) Part I plans. The wider context for this was the pressure from Brussels that the EIA (Environmental Impact Assessment) Directive might, as originally intended, be extended to include the environmental assessment of the plans and policies which give rise to development projects. The arguments for this strategic level of assessment were well aired: many agencies, both governmental bodies such as the NRA and non-governmental organisations such as the Royal Society for the Protection of Birds (RSPB), had been advocating the need to consider the cumulative effects of development, and secondary and indirect effects, as well as emphasising the advantages of evaluating possible impacts at the earliest stage of decision-making (Therivel *et al.* 1992).

The UK government, with its commitment to deregulation, and a political desire to stem, and if possible repeal, the flow of legislation from Brussels, had, as a gesture of prior compliance, published a guide to *Policy Appraisal and the Environment* (DoE 1991a), for central government civil servants. Although PPG 12 refers to this guidance as a help to LPAs in systematically appraising policies as part of the plan preparation process, practice amongst LPAs was in fact ahead of the advice. A number of local authorities had already followed up environmental audits of their areas with an appraisal of existing structure plans as part of the review process, such as Lancashire's ground-breaking work in 1991–92 (Pinfield 1992).

Lancashire had adapted the matrix approach of *Policy Appraisal and the Environment*, with the assessment undertaken using in-house expertise. Others, such as Kent County Council, had realised that there was a clear need to take a more strategic view of the large number of development proposals for which project-level environmental assessments were received under the Town and Country Environmental Assessment Regulations. As a first step towards this strategic view, Kent had again used a matrix approach to compare its existing structure plan with the emerging policies of the third review (Kent CC 1993). Other authorities had taken a very different approach: Ealing Borough, for instance, which had a tradition of planning for environmental quality and using the plan-making process as a forum rather than an end in itself, had tried to evaluate policies in its draft UDP in relation to the public's environmental goals, testing them in the community with different groups such as local people, developers and other public bodies (Wilson 1994).

The practice of environmental appraisal

It is quite clear that practice in some of the larger county and metropolitan planning authorities was evidence of a real capacity for innovation in incorporating the environment into land-use planning. Since the publication of PPG 12, many authorities have seriously considered and tried out a range of different approaches to appraisal, reflecting the different stages of their plan-making, the different characteristics of their areas, and their different planning functions.

Understandably, initially, LPAs raised a number of practical issues on how to undertake the appraisal. The most important related to the plan-formulation process itself: as plans have moved away from the survey–analysis–plan model of the early years of the current development plan system to a more complex, iterative process of goal-setting, policy formulation, implementation, monitoring and review, it was not always clear how environmental appraisal would fit in with policy formulation. While some, such as the District Planning Officers' Society, expressed reservations about the apparent primacy which environmental appraisal appeared to give to environmental objectives in the plan formulation process, it was generally accepted that there was a case for both undertaking the environmental appraisal of emerging policies and giving more weight to environmental considerations in the formulation of those policies (Wilson 1994).

Another, unresolved, issue relates to the timing of the appraisal, and its integration with public participation on the draft plan policies. District-wide plans were being drafted for all districts across the whole country, and some were about to go to Inquiry, and authorities were reluctant to open up new issues at this advanced stage. Others, while in their early stages, were reluctant to contemplate further delays through adding further complex appraisal or consultation stages; and undoubtedly many authorities felt they had neither the resources not the expertise to undertake the appraisal in-house or with consultants. Information about the baseline environmental conditions might be lacking, especially for those authorities without benefit of an environmental audit; but, as crucially, doubts were expressed about the ability to make predictions about the impact of different policies on the environment. This raises very real questions about the ability of the planning system and the planning profession, with its traditional lack of attention to the natural sciences in the educational curriculum, to address basic issues about the impacts of the outcomes of planning on the natural environment and to make judgements about their significance.

Developing methodologies

This gap in expertise has been overcome in a number of ways. One solution has been to use external sources of knowledge and experience in the appraisal itself, either through the use of consultants (such as Gordon District Council (Tyldesley 1995)) or, more commonly, through the inclusion on a panel of inputs from in-house, expertise from the local authority or another tier of local government, or from outside specialists (such as Oxfordshire 1996).

A number of local planning authorities were well down the road of policy-level appraisal before the Department of the Environment published its Good Practice Guide (DoE 1993) in the year following the publication of PPG 12. The Guide was based on the experience of a range of key authorities, and strongly influenced by the Kent model; at the time of the research which led to that guidance, it was found that already 'roughly one-fifth of authorities had embarked on environmental appraisal' (Barton 1994), quite a remarkable number given the above constraints. Only a couple of years later, a substantial proportion of LPAs have taken up the practice of environmental appraisal, including not only unitary and county councils, and district councils, but also regional bodies and conferences such as the West Midlands Regional Forum and the London Planning Advisory Committee (LPAC) (Therivel 1995). In her survey in late 1995, Therivel found that some three-quarters of English and Welsh LPAs were undertaking an appraisal (Therivel 1996). Figure 2.4 shows the status of EAs in England and Wales in late 1995.

There has been an extensive sharing of the variety of experience with developing methodologies among LPAs. While many have broadly followed an adapted version of the DoE's guidance in undertaking policy compatibility checks, and appraising policies against criteria for environmental stock, characterised as global, natural resources, and local environmental quality (DoE 1993), others, such as Hertfordshire, have adopted an approach which focuses more on the involvement of the public in the formulation of overall sustainability aims for the plan (Hertfordshire 1994). Others, such as LPAC and Greater Manchester, have established joint working to develop a common methodology, while authorities in Suffolk have jointly undertaken some of the stages such as characterising the local environmental stock (Therivel 1995).

Although the DoE's guide suggested key tasks for the appraisal, it was not intended to be prescriptive (Simon 1995), and Bedfordshire, for instance, in association with the RSPB, has published guidance based on

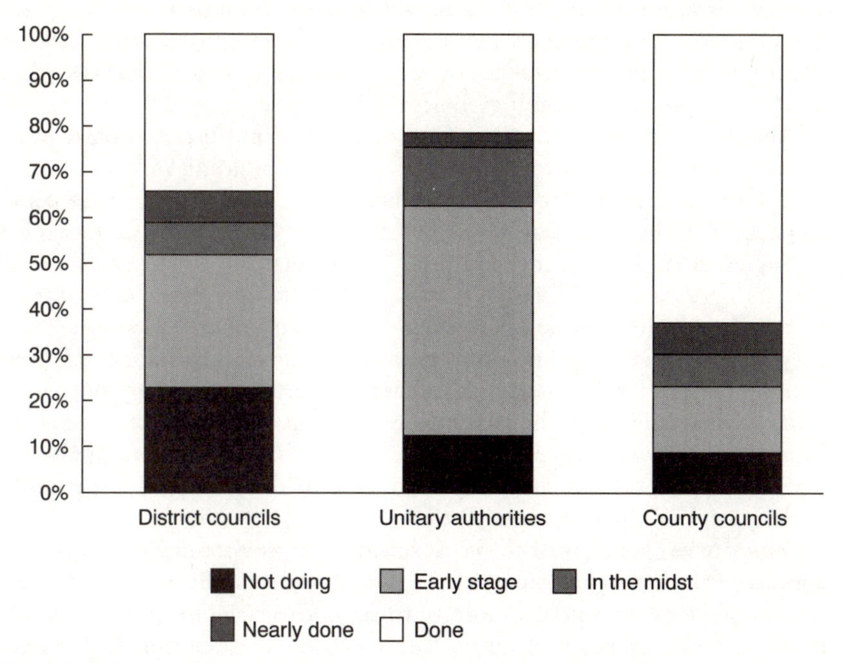

Figure 2.4 Status of local authority environmental assessments in late 1995
Source: Therivel 1996

its own experience with an iterative approach, revising the appraised plan in the light of the findings, and then appraising the revised plan (Bedfordshire/RSPB 1996). Some of the earlier appraisals used a fairly crude approach to aggregating the results of their scoring systems: the temptation in summarising the scores, whether they were based on numeric scores, symbols or ticks and crosses, was to aggregate in a way which implied equal weighting for the different criteria (Merrett 1994). More recent attempts have emphasised the importance of a commentary explaining the completion of the matrix, and, in particular, the judgements about whether impacts were highly or marginally significant, short term or long term and so on (e.g. Oxfordshire 1996).

Expertise and research

Even where procedural methodologies for policy appraisal are developed, however, it has to be recognised that the science of prediction is notoriously inexact, and that environmental interactions are complex, with unexpected secondary and indirect effects and unforeseen thresholds. Nevertheless, there are some areas, such as the complex relationship between land uses, transport, energy consumption and emissions where work commissioned by NGOs (e.g. Owens (1991) for the CPRE) or by government (e.g. Ecotec for DoE/DoT (1993c)), or undertaken by academics (e.g. Curtis and Headicar 1994), is gradually informing choices more explicitly.

Another source of guidance is the explosion of publications on relevant techniques for undertaking the appraisal and for building in more explicit stages in characterising the environmental stock in order to appraise policy impacts. For instance, statutory agencies and NGOs, such as the NRA (NRATR 1995), English Nature (English Nature 1994), and CPRE (Jacobs 1993), have advocated an approach which establishes critical natural capital (that is, assets of such value, rarity or irreplaceability that they must be given absolute protection), and constant environmental assets (where the sum total of assets should not be allowed to decline, but where some substitution might be possible). Some local planning authorities have been working with these concepts, although they are contentious – the RSPB, for instance, is concerned that over-defining critical sites might undermine less obvious assets (Buckley 1995).

Some have gone further in adopting an environmental capacities approach to identifying where predicted impacts might pose most challenges to the environment's capacity to support development or absorb its impacts, and a number of authorities (such as West Sussex 1995) have attempted to develop their appraisal in the framework of an explicit estimate of the carrying capacity of the authority's area. The implications of this approach are likely to permeate the whole plan: in the introduction to the Brecon Beacons Local Plan, for instance, linking the aims of the Plan to the statutory purposes of national parks, the document explains sustainable development as meaning 'development within the capacity of the environment. This may include reducing the impact of development; reusing and recycling land and materials; and managing the wise use of resources. This will require a more integrated approach to all land use planning' (Brecon Beacons 1994).

While many have reservations about the appropriateness of the concept of environmental capacity in the planning system (Barton 1995; Grigson 1995), the Department of the Environment has commissioned research into its feasibility (Entec 1996), and indeed has itself used the concept, for instance in the Sustainable Development Strategy (HMG 1994b). In a further adaptation of the concept, Friends of the Earth UK is promoting the idea of environmental space, which links the capacity approach with that of the ecological footprints of development (McLaren 1996).

Wales and Scotland

Local planning authorities in Wales might have been expected to be less enthusiastic in their attitude to the new area of environmental appraisal.

While Wales is covered by its own version of PPG 12, initiatives might have been substantially inhibited by the generally minimalist tenor of guidance from the Welsh Office, and the absence of other guidance encouraging consideration of sustainability objectives – such as the equivalent of PPG 13 on transport, which was only issued in 1996 in a form incorporated in the general PPG for Wales (Welsh Office 1996). Nevertheless, a number of both counties and districts, such as Powys County Council (1994) and Dinefwr Borough Council (1995), had undertaken environmental appraisal before the local government reorganisation into unitary authorities.

Practice in Scotland has been slow to follow that in England and Wales, with the appropriate guidance not being issued until 1996 in the Planning Advice Note on Local Plans (Scottish Office 1996). The lack of prior take-up by Regional Councils may have been partly attributable to the diversion of their attention from such strategic appraisal to the more immediate need to get their structure plans in place, before the abolition of the regional tier under local government reorganisation in 1996. However, a more precise methodology for appraising development plans was jointly commissioned by Gordon District Council, Scottish Natural Heritage and the Scottish Office, and tested on the Gordon District Consultative Draft Local Plan (Tyldesley 1995). This might be seen as an illustration of the closer professional working practices of central government and the local authorities in Scotland; alternatively, it might have been a gesture of support for the district tier in their new responsibilities.

In their report on the Gordon plan, the consultants appointed concluded that an essential initial stage of any appraisal was the drawing up of an appropriate set of environmental aims, in order to test the compatibility of the aims of the plan under review. The stages recommended lead from this initial compatibility consideration, through appraisal of the overall strategy of the plan, scoping of the issues addressed in the plan against a range of possible topics, to appraisal of the individual policies and proposals. The process should conclude with suggestions for monitoring and recommendations for changes. In the case of the Gordon plan, the consultants concluded that, while overall the plan's aims were compatible with environmental sustainability, and the scope of its policies was broad, there were problems with some of the policies, and in particular with specific proposals. '[T]he overall total of adverse effects is a cause of concern . . . The appraisal serves to indicate quite clearly that the level of development required in each settlement by the Structure Plan cannot be accommodated without significant damage to the environmental framework. The options for manoeuvre within the framework set by the Structure Plan are few' (Tyldesley 1995: para. 1.6).

The outcomes of environmental appraisal

This major critique of the constraints on the plan raises interesting questions about the outcomes of the appraisal process. Therivel notes that LPAs find presenting the results of the appraisal a challenge: 'All too often, the end-result of a policy appraisal is a large pile of undigested matrices' (Therivel 1995: 18). Some authorities, for instance, simply make a brief statement that an environmental assessment has been undertaken, and others, while including the EA in the local plan, do not explain how the appraisal has influenced the formulation of policies in the plan. However, more recent practice has been to go much further in explaining the process: Bedfordshire (1995) and Oxfordshire (1996) explain the changes made to the draft policies, while Chester City sets out the main conclusions of the appraisal at the end of each chapter in its local plan (Chester 1996).

How much change to the plans does the process of EA make? The consequence of the appraisals is for policies and proposals to be modified or reviewed where there is explicit recognition of the risk of potentially adverse effects, for firmer policies to mitigate these effects, and for new policies to be included. In some cases, the overall thrust of the plan has been changed by the appraisal, with more focus on the need to allocate land for development to reduce the need to travel, or to protect locally valued nature conservation assets. Transport and locational polices give considerable cause for concern: Oxfordshire, with its long-standing commitment to a green belt around the city of Oxford, and the promotion of development in the country towns, accepts that these policies 'would increase energy use from travel and transport' in the short term, before other policies to reduce the need to travel take effect (Oxfordshire 1996: 4 para. 23). Bedfordshire makes the point that 'The overall impact [of transport infrastructure policies] could potentially undermine the aim of contributing to sustainable development' (Bedfordshire 1995: 90 para. 4.19), but the appraisal has 'reinforced the need for policies which encourage an increased use of public transport, cycling and walking, and concentrate development in urban areas, thereby reducing the need to travel' (para. 4.22).

In other cases, the appraisal gives rise to radical new policies, such as Oxfordshire's policy EG3 requiring proposals for new energy-generating capacity to demonstrate that energy-conservation alternatives have been considered (Oxfordshire 1996). Although it is too early yet to draw definite conclusions on how far EA has altered the outcomes of the implementation of the plans, there is no doubt that the policies themselves are being changed by the process.

Sustainability appraisal

Just as the appraisal process has been used by local authorities to develop new techniques for identifying impacts on environmental stock through using capacity studies, so, independently of any central advice, many local authorities are taking the appraisal process much further into sustainability appraisals (such as Leicester City Council 1996). In this approach, the policies and proposals of the plan are assessed against locally generated and owned criteria expressing concerns for equity, distributional consequences, accessibility and quality of life, as well as local environmental quality. In many cases, these issues have derived from the authorities' commitment to the community-driven Local Agenda 21 process (for example, in Hertfordshire (Rumble 1995)), with much more explicit attention to both the social and the economic impacts of environmental polices, and to their potential for bringing about urban and social regeneration.

In addition, environmental appraisals are being undertaken not just of district plans, but of the highly contentious waste and minerals local plans. These appraisals will need to build in the commitment to the principles of proximity and regional self-sufficiency, explained above, and again are likely to lead to new methods for gaining consensus, such as that undertaken for waste management in Hampshire (Petts 1995).

Local authorities are also urged in the RSPB guide (Wilson 1993) to use similar methods to appraise the full range of their plans and polices such as Transport Policies and Programmes, Single Regeneration Budget Bids, and Economic Development Plans. At the same time, bids for European Union monies under the Structural Funds require a form of environmental appraisal (Wilson 1993): following the early efforts of the Strathclyde Partnership (Burleigh 1993), by 1996 all of the first tranche of these bids had been subjected to environmental appraisal, although the quality of the practice varies widely, and there is a case for best practice guidance (Seamark 1996).

In conclusion, the planning profession has probably surprised itself by its capacity to give serious attention to new mechanisms for integrating environmental concerns into plan-making. Just as central government seemed to have underestimated the likely number of environmental statements which would be submitted under the EIA legislation (with numbers averaging 300 a year in the early 1990s (DoE 1996a), compared with a prior estimate of twenty Annex I projects a year (McDonic 1988)), so the development of radical techniques associated with the take-up of environmental appraisal had been unforeseen. The extent of experimentation with radical new principles, policy instruments and techniques has been remarkable.

The role of the profession

Some interesting questions can be asked about the sources of this innovation, and how far the planning profession itself sought out a new justification for its existence after a decade of retrenchment in the 1980s, and in the face of the struggle to innovate and maintain any profile in the Compulsory Competitive Tendering (CCT) culture of the 1990s. While 'planners must be grateful that after the cold 1980s an issue has appeared which can give them some sort of legitimacy' (Marshall 1994), it is clear that the profession itself in the form of the professional body representing the interests of planning in the UK, the RTPI, has failed to take the role of leader or even advocate in this new arena. The profession undoubtedly did experience a crisis of confidence in this period, and is still concerned about the decline in take-up of student membership, even while RTPI-validated courses continue to increase (Discipline Network 1996).

The flourishing of environmentalism in planning has happened in spite of the Institute, which has consistently shown its reluctance to enlarge the scope of its interests. An example is its failure to establish a section of the membership representing the new profession of environmental assessment, despite the fact that the bulk of this new area of legislation lies within planning. In the absence of an open-door policy, a new body, the Institute of Environmental Assessment, has been established outside the RTPI, with its own claims to professional validation and review.

In the vacuum of leadership from the professional body, there has been scope for other influences on the shift in the focus of planning in the 1990s. Two important sources of influence which have been relatively neglected in other commentaries are the environmental pressure groups, and newly elected local authority members. The environmental pressure groups adopted a pro-active, chivvying role in defending planning, in commissioning research and in promoting best practice. The RSPB, for instance, has undertaken original research on the treatment of the environment in development plans (such as their *PlanScan* series (Bain *et al.* 1990; Davies *et al.* 1992; Dodd and Pritchard 1993); the Society commissioned and promoted key work on *Strategic Environmental Assessment* (Therivel *et al.* 1992); and has published best-practice guides jointly with local authorities (such as that with Bedfordshire on the environmental appraisal of development plans previously mentioned). Friends of the Earth and the CPRE have undertaken similar roles.

Local political leadership has also played a significant role. This has not generally been in response to the election of Green Party councillors (whose showing in local government, as with general elections, has been abysmal since the high point of their 15 per cent of the UK vote in the 1989 elections to the European Parliament), but to other features of local government. In Bristol, the London Borough of Sutton, Oxfordshire,

Kirklees, and Lancashire, the push for environmental audits and statements came from members, in some cases responding to pressure-group initiatives, such as Friends of the Earth's *Charter for Local Government* (McLaren and Adams 1989), or to the breaking up of the traditional bipartite dominance of committees. Members tended not to be so bound by professional or departmental boundaries in the examination of new areas, and were perhaps keen to have tangible evidence of a new field for local authority activity.

While local government has of course been subject to constant change, with the imposition of stringent budget constraints, performance indicators and CCT, the environment may have been seen as a source of potential local initiative and pride.

Conclusions

This review of the adaptation of land-use planning to the environmental agenda has shown the variety of practice that has developed in the 1990s, and the possible explanations for this variety. It is tempting, perhaps to satisfy our need for orderly classification of causes, to suggest that this variety is evidence of the continuing ability of the local level to adapt and innovate to protect its own interests, be they defined as private or public interests, and of the importance of recognising the many gaps between the expression of centrally intended policy outcomes and their implementation. What I have tried to show is that the forces accounting for this change are themselves very varied and in many ways contradictory, and that it should not be surprising that the pattern of response is complex.

Nevertheless, there seem to be some firm principles of sustainability being established which are influencing both local planning authority activities, and those of central government. These principles can be summarised as:

- carrying capacity
- critical natural capital and constant natural assets
- demand management, and the wise use of non-renewable resources
- the precautionary approach
- the proximity principle
- environmental enhancement
- local accountability
- quality of life.

As this chapter has shown, some of these are the initiatives of non-governmental agencies; some emanate from local planning authorities themselves, and some (such as the proximity principle) seem to emerge from central government. They have emerged in response to attempts by the centre, responding to the familiar neo-liberal agenda of the New Right, to establish a general framework within which individuals and businesses can accept their environmental responsibilities, at the same time as the professedly more traditional view of the importance of communities, including urban communities, has promoted the value of quality of place. They can also be seen as the outcome of internationally agreed conventions, where implementation has been seen to be important as a diplomatic tool, and at the same time as the outcome of perceptions at the local level that, in the context of a deregulatory government, action was needed to respond to anxieties about the quality of life.

They represent a much more complex and politically charged set of principles than the traditional preoccupations of planning with amenity, balance and self-containment, or even community development. Nevertheless, despite having to work with a still discretionary land-use planning system which has consistently eschewed principles in favour of pragmatism, the policy measures and instruments needed to implement these principles are being widely adopted.

These include, for instance, the formulation of policies to protect critical assets such as water resources and local air quality, and to conserve biodiversity at a level greater than just designated sites or for named species; and the array of policies and initiatives to minimise the need to travel, and to encourage more environmentally benign forms of travel. Of critical importance, too, is the way Local Planning Authorities are turning to the new measures available to them to give effect to some of these intentions: to use environmental appraisal and sustainability appraisal in a creative way to influence both plan preparation and the cumulative impacts of its implementation; to integrate the work on environmental indicators and targets into plan preparation and policy implementation and development control monitoring, and as a means of gaining local involvement in the process; and the creative use of planning conditions and obligations to achieve some of the principles such as environmental enhancement.

This attention to both the plan preparation process *and* to the outcomes of planning suggest that the integration of environmental concerns into planning is a secure development which will be maintained and will deliver some fundamentally different outcomes in the next decade.

3

CONSERVATION OF THE BUILT ENVIRONMENT UNDER THE CONSERVATIVES

Peter J. Larkham and Heather Barrett

> In 1979, the Queen's Speech contained a ringing commitment to 'bring forward measures to protect our national heritage of historic buildings and artistic treasures'. What is unforgivable, a decade later, is that ministers should have quite forgotten these words, even though membership of the National Trust exceeds two million, showing that more people care for the preservation of our heritage then ever before.
>
> (Binney 1991)

Introduction

Continuity, consolidation or change?

Conservation is an elusive subject to address in an analysis of political influences in planning. To a considerable extent the key trend has been continuity. Legislative and administrative systems for conserving the built environment had been put in place long before 1979, operating on both national and local scales. Key high-profile heritage events, which served both to raise consciousness of the importance of conservation and to shape future approaches, had already happened: including European Architectural Heritage Year in 1975, and the shock sale of Mentmore Towers in 1977 (see House of Commons (1978) for a critical commentary on this key event). Conservation since 1979 simply continued along lines already laid down.

Yet there have been developments in these systems. The Mentmore sale provoked an inquiry by the House of Commons Expenditure Committee and a White Paper by the outgoing Labour administration. The new Conservative government then 'moved surprisingly quickly. By the Spring of 1980 a new Act reached the statute book setting up an entirely new body, the National Heritage Memorial Fund, with independent trustees appointed by the Prime Minister' (Gaze 1988: 155). The National Heritage Act of 1983 set up a new quango, English Heritage (EH), with a wide range of heritage responsibilities. In the periodic review of planning legislation, conservation emerged as a distinct legal issue with the 1990 Planning (Listed Buildings and Conservation Areas) Act separate from the 1990 Town and Country Planning Act – although this new Act was largely consolidating existing practices and systems, not instituting new directions. Government planning policy has been explained in a series of Planning Policy Guidance Notes (PPGs), more concise and accessible than Circulars from government departments. Conservation has been no exception, with PPG 15 appearing – after a long gestation and consultation period – in 1994 (DoE 1994d). This did cover some new ground, for example in its explicit mention of 'the wider historic landscape', its treatment of economic issues, and its linking of transport with conservation.

But there have been new directions. An entirely new government department has been created, the Department of National Heritage (DNH), taking responsibility for built-environment conservation policy from the Department of the Environment (DoE). The importance of privatisation and market forces, so evident in other aspects of Conservatism, are also evident in the changing role of EH into the 1990s, and the urge to seek funding for conservation from business and community sources, or from the new National Lottery.

So the overall direction of conservation under the Conservatives is difficult to elucidate. Certainly there have been notable increases in the numbers of listed buildings and conservation areas. Heritage venues remain as key tourist attractions and income generators. The National Trust, a major non-governmental heritage organisation, celebrated its centenary amidst great public fanfare. The media profile of conservation remains high, with a new dedicated television series, *One Foot in the Past*. Yet much concern has been expressed, in the public media and professional press, over the direction that conservation is taking. Is Britain now dominated by the 'heritage industry'? Do we have too many conservation areas and listed buildings? Are we in danger of leaving important heritage issues, and even monuments, to the mercy of market forces? These and related questions form the focus of this chapter, which examines the emergence of key issues and their influence at national and local scales. Since much of conservation activity is, by its very nature, a local activity, we draw together these issues through examining micro-scale activities in two city-centre conservation areas.

Thatcherism, the New Right, history and conservation

Nigel Lawson (1992) has described Thatcherism as a mixture of free markets, financial discipline, firm control over public expenditure, tax cuts, nationalism, 'Victorian values', privatisation and a dash of populism. Thatcherism was a product of a generation of apparent national decline, and it could be argued that its central purpose was to make Britain great again. In so doing, it used history and heritage. 'The alleged authority of history is a useful tool in the construction of a new set of values: the past – however distorted – supplies a framework of justification for the present' (Hewison 1995: 211).

Yet Thatcherism was instinctive, rather than philosophical; and contained inherent contradictions between a free economy and a strong central state (Hewison 1995: 213). The numerically small New Right, a very diverse grouping, became influential and promoted a new consensus around many of the views of Thatcherism (Hewison 1995: 214–16). The results of this Thatcher/New Right coalition, and indeed its contradictions, can readily be seen at work in the field of conservation and heritage, as this chapter explores.

Some conservation history

Vital to an understanding of UK conservation is the endurance of its major organisations and systems. The National Trust, for example, was founded in 1895. Its early focus was on the acquisition and preservation, on the nation's behalf and in perpetuity, of places of scenic beauty. From the mid-1930s, as tax regimes changed to disadvantage the traditionally wealthy landed gentry, the Trust began to acquire country houses and estates. Although some families remained in residence, the intention was to promote public access. Its most recent acquisitions have included an Edwardian semi-detached house, and a Modernist 1930s house in London. The Trust is now the largest private landowner in the UK, owning some 235,000 ha of land and 756 km of coastline, 190 historic houses and castles, and 130 other properties (House of Commons 1994: viii). Its changing approaches have been well chronicled as part of its centenary celebrations (e.g. Jenkins and James 1994); however, one acerbic critique maintains that this has always been, and remains, an élite organisation dedicated to maintaining the property of the élite at the expense of the middle-class visitor and general taxpayer (Weideger 1994). Similarly long-established voluntary bodies have come to be regarded as expert advisers, which are now by statute to be consulted on relevant planning matters. These include the Society for the Protection of Ancient Buildings (SPAB) (founded 1877), the Ancient Monuments Society (1921), the Georgian Group (1937), the Victorian Society (1958) and the Twentieth Century Society (originally founded in 1980 as the Thirties Society). Their dates of formation chronicle the changing acceptance of different architectural and planning periods as being conservation-worthy.

There are two key systems of built heritage conservation in the UK:

those of the listed building and the conservation area. The former arose out of awareness that bomb damage during the Second World War was destroying many fine, unrecorded, buildings of historic and architectural merit. An initial national system was set up during the war, formalised by the 1944 and 1947 Town and Country Planning Acts (Harvey 1993). It became a statutory duty of the relevant Minister to schedule buildings of importance. Currently some 500,000 buildings are listed. The Secretary of State for National Heritage has statutory responsibility for listing, advised by EH (similar systems operate in Scotland and Wales). Having completed both an initial country-wide survey, then a resurvey, EH is now proceeding by thematic inspection of particular building types (Robinson 1990).

The conservation area concept was influenced by legal decisions in the early 1960s and by the French system of *secteurs sauvegardes* introduced in 1962. Duncan Sandys MP, President of the Civic Trust, introduced the idea into the UK in the 1967 Civic Amenities Act. Like much conservationist legislation in the UK, this was a Private Member's Bill. These areas were defined as 'areas of special architectural or historic interest, the character or appearance of which it is desirable to preserve or enhance': a definition which remains unchanged today. Designation was a simple process largely carried out by the local planning authority (LPA). It is a popular process, with nearly 10,000 areas designated across the UK by 1996 (Larkham 1996).

Personalities have thus always played an important part in conservation, from William Morris's founding of the SPAB to the present. As has been said, individual Members of Parliament have been influential in promoting Private Member's Bills on heritage-related issues, for which the governments of the day would give no official parliamentary time. During the last two decades, Secretaries of State have apparently become more personally influential. Michael Heseltine, when first in office as Secretary of State for the Environment in the early 1980s, supported and resourced the accelerated national resurvey (Brunskill 1993: 78). He took a personal interest in heritage issues which had previously been relegated to junior ministers who, conservation groups felt, 'were seldom in the job long enough to get a grasp of the complex issues involved' (Andreae 1996: 150). However, Nicholas Ridley, half a decade later, acquired a considerable reputation as a 'pro-development' minister, decrying conservationists as 'NIMBY' (Not In My Back Yard). Reviewing Ridley's treatment of EH, Lord Kennet wrote that when Mr Heseltine hived off English Heritage from the Department of the Environment in the 1980s, he made a point of saying that this would make it more visible, thus making any mistreatment of it more visible, and would therefore add to its influence. He had reckoned without Mr Ridley, who appears not to have cared whether his mistreatment of it was visible or not (Kennet 1991).

Ridley produced a well-known and oft-cited general attack on conservation, making many of the Conservative points on individual freedom and anti-bureaucratic control:

> I have a recurring nightmare, that sometime in the next century the entire country will be designated under some conservation order or another. The people actually living there will be smothered with bureaucratic instructions limiting their freedom. We will have created a sanitised, bureaucratised and ossified countryside out of something which has always been, and should always be, a product of the interaction of man [*sic*] and his environment as time goes by.
>
> (Ridley, speech to the National Association of Conservative Graduates, quoted by Suddards 1988: 523)

Ridley also granted permission for the demolition of a series of listed buildings in a London conservation area, on the grounds that the proposed replacement building would be a 'possible masterpiece', being of such quality that it would contribute 'more both to the immediate environment and to the architectural heritage than the retention of the existing buildings' (Watson 1991; Bar-Hillel 1991). This case, No. 1 Poultry, involved another personality: Peter (now Lord) Palumbo, a wealthy property developer who had spent years in assembling this site, and several attempts to demolish and rebuild on it.

The No. 1 Poultry case was argued through the process of a planning application, then on Appeal to the Secretary of State, through the courts, and was finally decided in the House of Lords in 1991. Although five Law Lords accepted that Ridley was procedurally correct in granting permission, they did not endorse his views on the quality of the new building, and specifically stated that they regarded this ruling as an exception, rather than forming a precedent (Watson 1991; Bar-Hillel 1991). Yet this case must have implications for development in conservation areas and affecting listed buildings (Larkham 1995: 101) particularly when development pressure increases after the recession. This single case is symptomatic of a clear tendency throughout planning decision-making in the 1980s: that is, the rising tendency for developers to appeal to the Secretary of State against a refusal of permission from an LPA, or to take the case to the courts.[1] In conservation issues, the courts have been very significant: the *Steinberg* decision suggested that development in a conservation area should not merely do no harm to the character or appearance of the area, but should positively enhance it (see the legal definition quoted earlier). A later case accepted that 'neutral' development, which neither harmed nor enhanced the area, should also be allowable. These interpretations had significant impacts on the way in which LPAs sought to control development in conservation areas (cf. Millichap 1989, 1989a).

Calamity and the conservation debate

A final set of events which has driven recent conservation thinking has been the series of high-profile disasters occurring since 1979. These disasters have included the fires at York Minster (1984), Hampton Court (1986), Uppark House (1989) and Windsor Castle (1992). These have, without doubt, served to stimulate a number of debates regarding conservation, from the basic principles of why these buildings were conserved and for whom, and whether they should be restored at all, to the minutiae of how restoration should occur (Fishlock 1992). Of particular relevance has been the issue of who funds restoration. The Queen is contributing to the restoration of Windsor through revenues from opening Buckingham Palace to tourists. Uppark has been restored to its pre-fire glory through a peculiarity of the National Trust's insurance, which financed complete restoration but would not have allowed any other option. The State has very much taken a back seat in these cases.

Other 'disasters', however, are man-made; and it is worth recording the demolition of the Firestone factory in London. This, a 1930s Art Deco factory building on the Great North Road, was subject to a development proposal and was hastily demolished over the course of the August Bank Holiday 1980, before a decision on its spot-listing could be made. This case alone spurred the government – and Michael Heseltine – to significantly accelerate the national resurvey of listed buildings (Robertson 1993: 29).

Conservation versus heritage

Commodification and marketing

'Heritage' has become an increasingly significant term in the conservation/preservation debate from the early 1980s. In the UK, as elsewhere in the developed Western world, de-industrialisation has led to a growing reliance on service-sector industries, of which heritage tourism is very significant (Herbert 1995). Polemicists such as Hewison (1987) argue against the rise of 'heritage', the selectivity and sanitisation of the images of past places presented, and the dependency on a museum-based industry and culture.

'Heritage' is neither history nor place: it is a process. This process includes the selection and presentation of aspects of both history and place, for popular consumption. 'Heritage is history processed through mythology, ideology, nationalism, local pride, romantic ideas or just plain marketing, into a commodity' (Schouten 1995: 21). It is a form of commodification. Therefore, 'heritage' means something quite different from 'conserved relict historical resources', and Ashworth (1992) argues that selection is central to the process. Although Lowenthal (1985) suggests that we all individually interpret the past in some form, its management and interpretation as heritage bring problems: 'some of what now purports to be heritage has been antiqued, not only in appearance but, rather more sinisterly, in being presented as if it was significant historically as well as being ennobled by time' (Fowler 1989: 60). The concepts of conservation and heritage are thus quite separate, although in recent years there has been a tendency to confuse, if not conflate, them.

Heritage has become increasingly used in the marketing of products and, especially relevant in the current context, places – whether individual sites or monuments, or entire 'tourist–historic cities' (cf Ashworth and Tunbridge 1990). Regeneration efforts in several neglected urban quarters have used heritage as a key component of the place-marketing and revitalisation strategies, and this can clearly be seen in Bradford's Little Germany, Nottingham's Lace Market and Birmingham's Jewellery Quarter (Tiesdell *et al.* 1996). Yet criticism surrounds the selectivity of heritage as used in place-marketing, which inevitably excludes aspects of local heritage deemed 'unsaleable' to tourists or investors (Kearns and Philo 1993; Gold and Ward 1994). There is also increasing criticism of the sanitisation of heritage, which is clearly seen in the built environment in pedestrianisation, pseudo-historicist street furniture and enhancement schemes (Booth 1993). But such is the competitive nature of contemporary place-marketing that such questioning is seen as unwelcome, even traitorous (see, for example, the political furore generated by the local ruling Labour group surrounding Loftman and Nevin's 1992 critical study of Birmingham's flagship regeneration projects).

The selectivity inherent in the heritage concept is especially problematic in a multi-cultural and/or historically diverse context, and Tunbridge (1994: 123) suggests that 'the political implications of culturally selective identification and interpretation, conservation and marketing of the inherited built environment are profound and potentially deadly'. He particularly addresses the question of 'dissonant heritage': the heritage of oppression and war is difficult to present without causing offence to some socio-political groups. Yet such heritages may be present in subtle, but nevertheless readily perceived, ways (see also Tunbridge 1995).

Selectivity: what is conservable?

Issues of selectivity inevitably permeate conservation. There are legal definitions of both conservation areas and listed buildings, although these are subject to interpretation by the relevant decision-makers. For example, over the last two decades, as the number of conservation areas has grown, the types of area deemed worthy of designation have changed markedly (Larkham 1997). From initial concentrations on mediaeval and Georgian towns, there are now cemeteries, canals, country houses, expanses of agricultural landscapes, parts of 1930s semi-detached suburbia, and the Settle–Carlisle railway. Whilst the earliest definitions were tightly defined parts of towns, many recent designations have been extremely large: the first of the two areas in Yorkshire protecting the landscape of dry-stone walls and field barns is of 72 km^2; and, after several extensions, the Bath conservation area covers 1,914 ha, or 66 per cent of the city's area.

One innovative designation was made by Birmingham City Council in 1988 and covered an area of speculative 1930s semi-detached houses. The issues surrounding this designation even led to a feature article in *The Times*:

> Birmingham's aim is to wrap this arcadia in aspic by means of a conservation order, and all the signs are that it will manage to do so . . . If it does, then this may be the very first development of the period to be thus protected in this country. In practical terms it would mean that nothing in the designated area of about 150 homes could be added to or altered unless strictly in the style of the original. . . . One immediate result of that would be to prevent any more of the bay windows disappearing as the softwood rots and the owners look for a cheaper replacement. . . . School Road happens to be an excellent example of the *genus*, and the whole city has become so sensitive about conserving what is good that it is now doing so long before a desperate rearguard action is required.
>
> (Franks 1988: 11)

The justification for designation, recalling that a conservation area must possess 'special' interest, is that this area is unusual in that it has remained largely unchanged since development: no infilling, few extensions or other alterations, and most buildings retain original period features. This was, perhaps, a brave designation, causing some debate locally and professionally over the intrinsic conservation-worthiness of such a 'typical' area.

Likewise, views on what is listable have changed over time. The gradual acceptance that Victorian and industrial buildings are important is well known; but the resistance to twentieth-century buildings has been considerable. Only since 1979 have inter-war buildings been listed. The debate about post-war buildings remains intense, with the Secretary of State accepting only seventeen of the first list of about seventy put forward by EH in 1988. A similar debate arose over protecting the familiar red telephone boxes, threatened by British Telecom's programme of modernisation. The reaction in some areas was to put these forward for listing, and some 2,000 have been listed: the question of why some, and not others, or indeed all, remains unanswered.

Selectivity also applies to the funding of conservation projects.

Obviously, not all can be grant-aided. Yet the reasons sometimes given for the refusal of assistance for a range of heritage projects can show a tortuous logic. A prominent example of the early 1990s, which received much adverse criticism, was EH's refusal to assist an appeal for £200,000 from the National Maritime Trust to fund a new lower foremast for the *Cutty Sark*, itself a Grade I listed building, on the grounds that masts were not essential because the ship formed a static display (*The Times,* 20 March: 6). Conversely, National Lottery funding has been refused for the restoration of the UK's last surviving record-breaking aircraft, apparently because the application was for restoration to airworthy condition rather than for static museum display.

Whose heritage?

There is a continuing concern that much conservation remains an élite activity and that, even where 'ordinary' or working-class issues become heritage and are conserved, they are over-sanitised. The nature of the debate over 'authenticity', particularly during the late 1980s and focusing on the large open-air industrial museums, typifies these concerns (West 1988).

The heritage bodies remain targets for similar criticisms of élitism. Weideger's view of the National Trust has become well known: and, although Delafon's critical review of her book suggests that it is weak on analysis and not well researched (Delafons 1995), it remains the only overview of the Trust by an outsider. Even the Trust's own histories cannot disguise the fact that it has largely been run by a particular social and educational élite for many decades (cf Gaze 1988; Jenkins and James 1994). There have been criticisms that the Trust's values are anti-democratic (Wright 1987), that it has a southern and middle-class bias in its acquisitions (Clark 1986), and that its approach to the heritage of the twentieth century is blinkered (Pearman 1992).

One of the key events touching upon the issue of 'whose heritage', indeed raising questions of privatising and denying admission to certain elements in society, was the Stonehenge solstice injunction in 1985. For several years, large groups of travellers had congregated around Stonehenge at the summer solstice, camping for weeks and holding pop festivals. There were 'allegations of all sorts of illegalities and undesirable activities, and undeniable mess and damage' (Gaze 1988: 84). Eventually, the National Trust (owner of the surrounding land) and the Department of the Environment (custodian of the monument) obtained a court injunction and, with a massive police presence in succeeding years, prevented access by this group. The stones are now fenced off for much of the year.

The rise of the market?

Market forces have become significant in many areas of conservation and heritage, not least in debates over the function of conserved areas and buildings – economic re-use and/or tourist attraction versus strict preservation. This has become particularly clear in the issues surrounding the disposal of large conservation-worthy areas and buildings formerly required for defence purposes, following the end of the Cold War (House of Commons 1995).

A further relevant aspect of the 'market forces' debate clearly evident in Conservative governmental policy is the intention to deregulate: to 'lift the burden' of governmental bureaucracy from business (DoE 1985a) and to ease the financial burden to central government of what had been accepted as government responsibilities. Both aspects are visible in conservation.

In terms of 'lifting the burden', the clear example is the introduction of the 'Certificate of Immunity from Listing' by Heseltine in 1980. This was a reaction to developers' protests that buildings purchased for redevelopment were promptly being listed, thus frustrating the development – or at least making it far more difficult and expensive. The key case cited as a motivation for this was the Johnny Walker whisky warehouse in Tower Hamlets: 'Bought for millions for redevelopment it was promptly listed (27 September 1973)' and a subsequent court judgement held that this was an acceptable commercial risk (Robertson 1993: 28–9). In recent years, developers have voiced similar protests over conservation area designations immediately following development proposals, particularly since this gives the LPA power to control all demolition within the designated area (cited in Jones and Larkham 1993).

Another aspect of deregulation was the proposal in EH's policy review *Managing England's Heritage: Setting our Priorities for the 1990s* (English Heritage 1992) to transfer management and financial responsibility for 203 monuments in its guardianship; to seek commercial partners to help fund its preservation programme; and to transfer responsibility 'for most decisions concerning the historic environment' in London to the individual London Boroughs. These proposals brought a storm of protest. Letters immediately appeared in *The Times* from Dr Peter Addyman, President of the Council for British Archaeology, Professor Lord Renfrew, Master of Jesus College, Cambridge and Henry Cleere, ICOMOS World Heritage Coordinator. The DoE informed EH that its proposals for devolving conservation powers in London were unacceptable on cost grounds (*London Evening Standard*, 13 April 1993: 14). The critic Brian Sewell attacked the plans 'to devolve responsibility for 15,000 Grade II listed buildings saying that Mr Stevens [EH Chairman] has left valuable properties to the "mercy" of local authorities unable to resist the determined efforts of developers' (*London Evening Standard*, 20 April 1993: 9).

Privatisation, another tenet of Conservative philosophy, has also raised issues for conservation. EH planned, in 1993, to privatise its Historic Properties Restoration Department and Design and Works Department by seeking tenders from contractors. These plans were strongly criticised, with suggestions that standards of conservation and repairs to monuments would be threatened (*The Times*, 3 April). Michael Portillo, then Chief Secretary to the Treasury, also instigated a 'review of antiquities' in 1994 which included examining ways in which the private sector could be involved in running monuments managed by EH: for example, 'buying leases to exploit the commercial potential of castles, historic houses and ancient monuments' (*The Times*, 6 August: 5).

However, the internal dichotomies of the government's approach to market forces have been identified by John Delafons (a former senior civil servant at the DoE), in a review of the first draft of PPG 15 and its accompanying consultation paper. He suggested that there were clear opposing views in the DoE and DNH.

> What is interesting about this avalanche of policy guidance is not the relatively minor proposals for changes to the development control regime (on most of which the consultation paper pours cold water) but the shifts in policy emphasis and the tension between the conservationist philosophy of the Department for National Heritage and the deregulatory doctrine that the Department of the Environment has evidently been obliged to assert. This results in some perilous see-saw drafting . . . If the deregulatory tendency predominates in the consultation paper, it is forced to take a back seat in the new draft PPG. . . . once it gets into its stride, the conservationist interest clearly predominates and the deregulatory imperative is tacked onto it like a scrap of graffiti on a listed building.
>
> (Delafons 1993: 226)

Tensions: central versus local states

The structure of the UK planning system, on which so much conservation depends, has many inbuilt tensions. Some tensions within different departments of central government have already been revealed. Particularly significant is that between the central state – the level of national government – and the local state. The role of the Secretary of State, in person, in taking heritage–related decisions both in deciding whether or not to accept suggestions from EH to list buildings, and in deciding planning appeals, emphasises the superiority of the national to the local state.

Several cases exemplify this assertion of power over the wishes of the local state, its elected representatives, local people and even the government's own heritage advisers. British Telecom applied for planning permission and Scheduled Ancient Monument Consent for a telecommunications mast on The Trundle, near Chichester. Both LPA and planning inspector recommended refusal. However, the Secretary of State disagreed, and

> took the view that the proper test to be applied was not limited to the effect the proposed development would have on the monument both visually and physically in terms of its effect on the archaeological remains, but should include the need to erect the mast and radio station . . . in short, in determining the application harm to cultural heritage has to be weighed against telecommunications needs.
>
> (*Journal of Planning and Environment Law* 1991: 301)

Lord Hesketh, the minister then responsible, delisted the Pump Room, Clifton, Bristol in 1991. This was only two years after it had been listed, and was against the advice of EH: the building's owners immediately applied for permission to demolish and redevelop (*The Times*, 10 January 1991).

The rise of the 'plan-led' planning system has also been problematic (Morton and Ayers 1993; Larkham 1994). Whilst ostensibly making LPAs more responsible for explicit policy, development plans are still subject to approval by the Secretary of State and he, or his inspectors, have made some odd recommendations. LPAs have attempted to produce comprehensive conservation guidance, yet on several occasions this has been deleted from the plan itself. In the case of Islington, all but two conservation area guidelines were removed from the plan itself: the Secretary of State feeling that,

> as they stood, the design policy guidelines could not justifiably be included in the plan. But that did not mean that part of the guidance could not be properly included, with the remainder being set out elsewhere as detailed guidance which did not have the same force and strength as that which was contained in the plan.
>
> (*Journal of Planning and Environment Law* 1995: 122)

The LPA felt so strongly that its detailed guidelines merited the statutory strength of inclusion in the plan that it challenged this ruling in court, but was unsuccessful.

One of the key problems about the government's decentralisation is that, although responsibilities and decision-making have nominally been decentralised (despite the above cases), finance has not followed. In archaeological conservation, for example, it was recently pointed out that 'in addition to the existing responsibility for archaeology within the planning system, LPAs will take on a significant financial and administrative burden if the government's recent proposals are implemented' (Pugh-Smith and Samuels 1996: 724).

The rise, role and fall of English Heritage

To a great extent, the history of conservation under the Conservatives is the history of English Heritage, some of which has already been discussed. This quango was established under the National Heritage Act 1983. Its duties include securing the preservation, and promoting the public's enjoyment and knowledge, of ancient monuments and buildings; and promoting the preservation and enhancement of conservation areas. EH manages over 400 properties directly, administers a range of conservation grants, and provides expert advice to government departments, LPAs and individuals. EH thus represents a centralisation of expertise, control and finance.

However, there have been suggestions that the division of responsibilities of EH and other agencies is illogical, and that there are potential conflicts of interest (House of Commons 1994: xix). Using EH Annual Reports as evidence, Andreae (formerly a senior officer with EH) has questioned whether EH had the sharp focus on identifying and dealing with priorities which its predecessor, the Historic Buildings Council, evidently did (Andreae 1996: 152).

A further perennial problem appears to have been repeated cuts in the real value of central government's grant to EH (e.g. Bond 1988). Although in 1995/6 some £41 million was offered in repair grants, including £12 million for secular buildings, this 'was a major reduction on the previous year's figure', and the new Conservation Area Partnership (CAP) scheme has been hit by new government spending targets:

> It had been English Heritage's intention to establish the round of bidding for new CAP schemes as an annual event, but after examining our spending commitments for the coming three to four years, we realised that there would be very little money available for new schemes to start in April 1997. In consequence, we were unfortunately forced to cancel the planned round of bidding . . . What is more, the amount now allocated nationally to these schemes leaves very little for any other work in conservation areas – for example direct buildings at risk grants or the funding of heritage-related environmental work.
>
> (Johnson 1996: 16)

Thus concentration of power has not been matched by resource allocation.

A review of media coverage also suggests that EH has a significant image problem. Most coverage was positive under the first chairman, the respected conservationist Lord Montagu of Beaulieu. However, his replacement by the former newspaper executive Jocelyn Stevens in 1991 marked a period of vehemently critical coverage. The conservationist Gavin Stamp suggested that he was 'an astonishing choice' (Stamp 1991). There were widespread reports of Stevens's approach, including one claim that he treated some visitors 'with discourtesy such as none of us had witnessed in our public and professional lives' (Anstey 1993). His acerbic reaction to criticism is frequently noted (e.g. Leedham 1993). However, EH's Chief Executive counter-claimed that there was a 'malicious' campaign by 'certain conservationists' against Stevens (*The Times*, 26 March: 17). Even if this were the case, one must ask why it should be so.

Symptomatic of EH's problems are the hostile reactions to its proposals to devolve responsibility for some monuments and otherwise privatise some services (English Heritage 1992: see earlier). Again the management style seems to be the problem.

> As described in their memorandum, English Heritage's proposals for local management might be thought to be reasonable. Why then did they arouse such a volume of hostile comment? Lack of consultation and the resultant surprise at the proposals were perhaps two factors. The brevity and lack of detail in the strategy document and the fact that the supplementary background paper which followed it was hardly more enlightening may have been other reasons.
>
> (House of Commons 1994: xxiii)

The Department of National Heritage

It is in the creation of the Department of National Heritage in 1992 that critics have seen the most direct Conservative intervention in this field. However the DNH's responsibilities include issues, and the management of 48 bodies which had previously been the responsibility of six departments (National Heritage Committee 1994: vi) together with new responsibilities including the National Lottery and Millennium Fund. Ravenscroft (1994: 134) notes the 'strong reference in the election manifesto to the department's orientation towards the private sector' and suggests that this pro-market drive could explain the 'curious lack of responsibility for countryside recreation and landscape heritage, both of which would appear to have been appropriate candidates for inclusion'.

Also curious is the division of responsibilities between DNH and DoE (DNH 1992). The new department now takes responsibility for conservation policy direction including the procedure for listing buildings, while the DoE retains all planning functions, including deciding upon appeals for planning permission and listed building consent in conservation areas.

It has been suggested that the nature and extent of this division caused major delay in issuing the influential jointly published PPG 15 (Hugh Corner, DNH, pers. comm.). Whilst this separation of 'heritage' from 'planning' remains problematic, the House of Commons National Heritage Committee quickly discovered other anomalies in responsibilities, for example recommending that 'responsibility for all relevant aspects of heritage in England ought to lie very clearly indeed with the Department of National Heritage', and that 'the funding of the Ministry of Defence's museums ought to be transferred to the responsibility of the Department of National Heritage' (House of Commons 1994: para. 62).

Ravenscroft's analysis of the DNH suggests that, in the terms of Thornley (1993), it reflects 'authoritarian decentrism': a centralised policy control mechanism used to promote market processes and, also, Major's 'classless society' rhetoric.

> Encouragement for the extension of domestic leisure opportunities in the private sector, together with a reduction of public control at the local level, through the imposition of the compulsory tendering of management contracts has, ostensibly, freed individuals from the economic constraint and servitude of social democracy. This has been replaced by a system where good, industrious, citizenship is rewarded not only by access to a wider and better range of leisure and tourism opportunities, but also, by association, to assimilation with the values of the prime minister and his government: values based on inheritance, individual freedom and a pride in the nation.
>
> (Ravenscroft 1994: 136)

A more acerbic critique has been given by Amery (1995), who criticises the slowness of action, suggests that the small size of the DNH in the government machinery causes problems, and is concerned that 'the future of the nation's heritage lies in the hands of this maverick bunch of officials, politicians and their quangos' (Amery 1995: 28). The DNH is also caricatured as the Department of Nothing Happening (Hewison 1993), a description which, in the authors' experience, has stuck in many local and national conservation agencies.

Conservation and practice

The locality debate and conservation

The issue of 'locality' has become a key theme in social science during the 1980s and, although there have been lengthy debates over definitions and usages (cf Cooke 1989; Duncan and Goodwin 1988), it clearly has some relevance to the very local-level concerns and battles over conservation. The concept recognises that localities are significant and unique, representing 'at any given time a synthesis of political, cultural, social and economic histories and characteristics. These locally based, historically produced characteristics have a dialectical relationship with wider social processes' (Meegan 1993: 59). That being so, and following a major ESRC research initiative, 'policies are mediated through . . . the particular socio-economic, political and cultural characteristics of place. Given that these characteristics are historically contingent and vary between places, policies can potentially operate with very different effects in different areas' (Meegan 1993: 59). Even so, some issues arising in the 1980s and 1990s raise fundamental politically related questions of competitiveness and power relations in conservation across, not just the UK, but Europe.

It is important to reinforce the local nature of much conservation, and thus its clear relevance to this debate. Even the definition of conservation areas and the 'special interest' of their 'character or appearance' shows this; and the bulk of conservation battles are fought at local scales. Despite their media profiles, there are relatively few battles for Euston Arch.

Locally, therefore, there is the question of who assesses the character and appearance of the 10,000 or so conservation areas. Indeed, Morton (1991) has argued that only perhaps 10 per cent of areas do have such appraisals! The example of Stratford upon Avon highlights these problems. This district has some eighty conservation areas, which are being reviewed by teams of consultants selected by competitive tender. Yet their differing approaches leads to problems of reconciling actions and proposed actions (reported in Jones and Larkham 1993). Character and appearance are locally distinctive and do not seem to be hard data susceptible to pseudo-scientific analysis.

A wider problem has been the intervention of EU legislation relating to free trade and competitiveness. For example, enhancement contracts above a given financial threshold should be advertised for tender throughout the EU. There is also the suggestion that conditions requiring the use of local materials should be prohibited as anti-competitive. Hence major pedestrianisation schemes in Scotland have had to be carried out using Spanish granite sets and Italian craftsmen (K. Murray, Tibbalds Monro; reported in Jones and Larkham 1993); and conditions requiring the use of Welsh roofing slates in Caernarfon have been queried.

Conservation in practice at local levels

Throughout the 1980s, conservation policy operation at the local level was undoubtedly affected by the tensions between conservation and the New Right at the national level, with contention extending into the 'everyday' negotiations between LPAs and developers. During this period, conservation became an increasingly important part of mainstream local authority planning practice, forming one of the key concerns around which a beleaguered profession sought to reinvent and justify itself (Healey 1989; Goodchild 1990). The higher profile for conservation at the local level was evident in two ways. First, conservation concerns became increasingly visible within local planning policies, building on *ad hoc* conservation initiatives of the 1970s and reflecting wider social and professional developments in heritage consciousness. Secondly, the spatial extent of conservation control increased, through the locally determined designation of more conservation areas and through extensions to areas already designated. Larkham and Jones noted that 'by far the majority of [conservation area] policies were first introduced by authorities in the 1980s. This corresponds with the high level of designations in this period and a peak in the building cycle' (Larkham and Jones 1993: 403).

The growing economic importance of heritage, and changing views on the 'conservation-worthiness' of nineteenth- and early twentieth-century buildings all undoubtedly fuelled the increasing number of area designations. However, it has also been suggested that the planning climate of the 1980s was influential in this trend (Punter 1990; Morton 1991; Larkham and Jones 1993). It is apparent that, as a locally designated and controlled planning tool, conservation areas offered the means by which LPAs could increase control over design issues and exert the leverage of demolition control, in the face of central government policy statements urging their reduced intervention in design issues (DoE 1980, 1985a; Punter 1986). It was, perhaps, the perceived historical ambiguity in the reasons behind the designation of many new areas that generated much of the tension between conservation and development at the local level. As the influence of conservation concerns grew, in defiance of the general deregulatory and enterprise planning rhetoric, developers increasingly moved to challenge local conservation controls. Nowhere were these problems more acutely expressed than in negotiations on developments within conservation areas covering the central business districts (CBDs) of cities, where conservation and business concerns intersected most clearly.

For LPAs attempting to operationalise conservation policies within CBDs, the impact of the emerging New Right planning ideology created two main problems which dominated conservation practice at the local level. First, local autonomy in conservation matters was frequently undermined by the granting on appeal to the Secretary of State of planning applications which had been refused at the local level on the basis of their detriment to local character, or which had not been determined owing to negotiation difficulties. Secondly, in an enterprise climate with reductions in state spending, LPAs found themselves increasingly compelled to seek private funding for conservation enhancement initiatives. Yet, while these problems were generally evident at the local level, the ability of LPAs to resist or modify the centralising, deregulatory and enterprise trends impacting on conservation practice varied. Key factors underpinning these variations in local influence included the local economic climate within which planning and conservation were carried out, the type of built fabric conserved, the availability of grant aid, the comprehensiveness of policies and the conservation expertise within the LPA.

These issues will be explored through a focus on conservation practice within two major English cities, Birmingham and Bristol, which experienced differing fortunes during the 1980s in the application of conservation policies in their cores (Barrett 1996). Discussion will concentrate specifically on developments occurring within conservation areas covering the CBDs of the two cities: the Colmore Row and Environs Conservation Area, Birmingham (designated in 1971) and the City and Queen Square Conservation Area, Bristol (designated in 1972) (figures 3.1 and 3.2).

Conservation in central Birmingham and Bristol

Within both areas, the 1980s was a period of increasing tension between conservation and redevelopment concerns, with rising commercial pressures exposing the weaknesses of non-statutory conservation controls. Both LPAs found it difficult to apply policies seeking to prevent character erosion and promote local distinctiveness as businesses sought to apply universal 'contextual' design solutions for new building and use standard 'corporate-heritage' elements for building interiors and exteriors in conservation areas (Barrett 1996). However, while the influence of national economic trends, planning policy development and architectural fashions produced a similar trajectory of conservation policy development in both areas during the 1980s, important local differences existed. Differences in the local development market, the extent of policy development in the two areas at the beginning of the 1980s and the extent of building listing produced contrasts in the 'success' of conservation polices in relation to development pressures. Of particular importance was the perceived worth of area fabric when judged against national criteria by developers and the DoE at appeal.

In common with trends evident at the national level, both the Birmingham and Bristol LPAs moved to strengthen and formalise earlier *ad hoc* conservation controls, bringing conservation further into mainstream planning practice. Both developed their first definitive conservation policy documents during the later 1980s (Birmingham City Council 1987; Bristol City Council 1989). The formal expression of conservation objectives proved timely for both areas in the face of a rising tide of development linked to increasing economic prosperity in the mid-1980s. Both LPAs also significantly increased the number of conservation areas in their respective cities, with the number in Birmingham rising from fifteen to twenty-five and in Bristol from sixteen to twenty-nine during the 1980s.

Figure 3.1 Colmore Row and Environs Conservation Area, Birmingham (redrawn from Ordnance Survey 1:1250 sheet: crown copyright reserved)

Figure 3.2 City and Queen Square Conservation Area, Bristol (redrawn from Ordnance Survey 1:1250 sheet: crown copyright reserved)

In addition, both LPAs extended a number of existing areas. The Birmingham LPA also developed a local list of historically important buildings not covered by national listing.

Between 1977 and 1987 in Bristol, long-term conservation programmes were supported by grants from the Historic Buildings Council (HBC) (now EH) following the granting of 'Priority Town' status in 1977. In order to maximise access to these funds, the coverage of conservation areas in the core was increased during this period, forming an interlocking grid of conservation control. In combination with a relatively buoyant demand for office accommodation (Bateman 1985) and the expansion of leisure uses around the Docks (Punter 1990), funding enabled the LPA to impose tight conservation controls on development. HBC funds enabled the LPA to initiate pump-priming enhancement projects, build up conservation expertise with the authority, and set high standards of conservation work for private developers to follow (Punter 1991). Within the centre of the city, planning became increasingly driven by conservation considerations, with conservation forming a key component of the 1990 Draft City Centre Local Plan (Bristol City Council 1990). However, progress during the second five-year programme between 1982 and 1987 proved problematic as local authority cash crises reduced the amount of required 'matched' funding from the City Council, and as developers sought to challenge the desirability of strong conservation controls.

In Birmingham, the development of conservation concerns was more tentative during the 1970s and 1980s, as the city gradually moved away from its post-war planning doctrine of comprehensive redevelopment and functional efficiency. Biting economic problems in the late 1970s and early 1980s (Spencer *et al.* 1986) generated a reluctance on the part of the LPA to impose any restrictions that would be seen to be deflecting business opportunities from the core (Birmingham Conservation Areas Advisory Committee (CAAC) 1984). Throughout the 1980s, much of Birmingham's conservation programme in the city core was forged from negotiation and compromise between the LPA and private business. With little access to national funds, such as HBC grants, and limited City Council funding, conservation initiatives remained largely driven by the private sector. As a result, the development of conservation enhancement initiatives proved problematic as developers and private businesses declined to incur extra costs. In addition, conservation efforts in the city were frustrated by the relatively low number of buildings protected by listing in a city with a predominantly post-1800 fabric. Although the LPA attempted to address this issue by developing the local list of important historic buildings, this initiative had limited impact owing to its lack of legal standing. Consequently, the LPA utilised conservation area designation and extension as a means to enhance local control.

Moves to extend the Colmore Row Conservation Area in the 1980s reflected this desire of the LPA to increase local controls in the face of national listing ambivalence in the core. In 1984, parts of Corporation Street and New Street containing the remaining nineteenth-century retail fabric were incorporated into the conservation area as part of proposals in the Central Area Local Plan (Birmingham City Council 1984) (figure 3.1). However, an extension to remove boundary anomalies to the north of Edmund Street was rejected on the basis of the number of modern office buildings in the area. Problems resulting from this fragmented boundary emerged immediately after this decision, when the replacement of a block of Victorian buildings at 160–170 Edmund Street by a tall office block comprising a three-storey podium surmounted by a seven-storey rectangular tower was proposed. The scheme was refused by the LPA, on the grounds that a tall tower would be detrimental to the listed 158 Edmund Street and the adjacent conservation area. However, as the buildings lay outside the conservation area boundary, the LPA lacked control over their demolition. The LPA and the amenity societies therefore sought the listing of the building group, although this was rejected by the DoE which did not consider the buildings to be of sufficient merit. The LPA then sought to use the locally determined option of extending the conservation area to include the Edmund Street buildings, in order to obtain control. This was approved by the planning committee on 25 July 1985, despite objections against this localised imposition of control from the block's landowners. They argued that the buildings on the site were of no architectural interest and that designation prejudiced future 'economic' development. There was, however, considerable local amenity and business support for the move, these groups viewing the extension of the conservation area as a means 'to protect Birmingham from the glass tower'. The use of conservation area designation to provide demolition control powers and enhance local negotiating strength in the face of national ambivalence was clearly apparent in this case.

Increasing central/local tensions: negotiation and appeals

It is clear that general planning pronouncements at the national scale had a significant impact on conservation management at the local level. While development control in conservation areas should have been independent of moves to streamline the planning process, with decisions more locally based, assertive developers were quick to put pressure on LPAs for proposals within commercial conservation areas. Tensions between national and local concerns were particularly reflected in the time taken to process applications and in the outcome of appeals to the DoE. Considering the average time taken to process applications in the two areas in the 1970s and 1980s, both Birmingham and Bristol showed a decrease in the number of days taken in the early 1980s (figure 3.3), reflecting the impact of

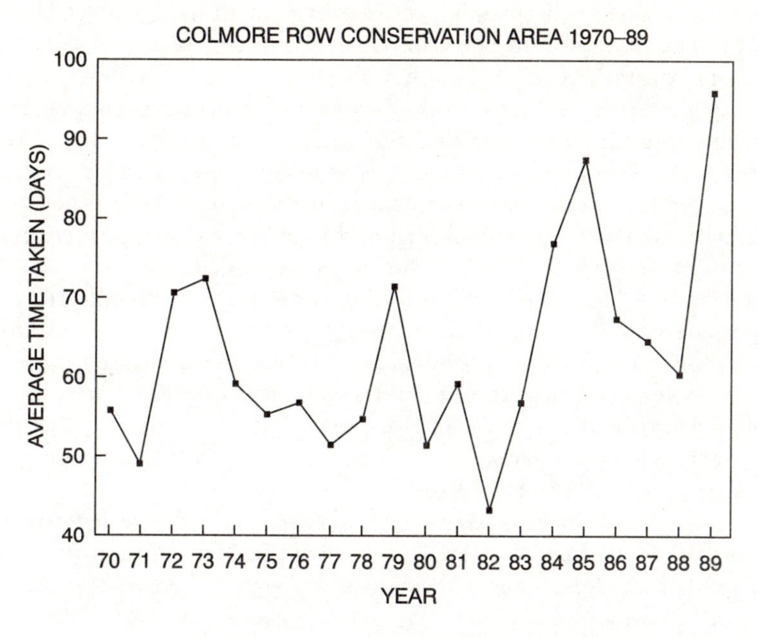

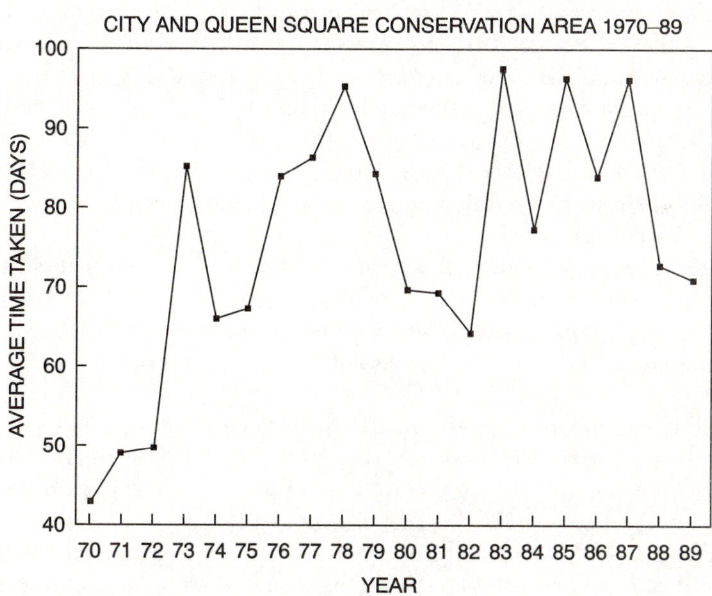

Figure 3.3 Time taken in determining planning applications in (top) Birmingham study area (bottom) Bristol study area

moves to streamline the planning process evident in national trends (DoE 1991).[2] The dramatic increase in time taken in the mid-1980s, also mirroring national trends, highlights the problem faced by LPAs in dealing with increased application submissions whilst maintaining a commitment to conservation control and negotiation; this precipitated increased conflict with central government and development interests. In Bristol, the more erratic levels of average time taken in the mid-1980s can be linked principally to the wider attacks launched by development interests on conservation controls in the city. During this period, considerable pressure was put on the LPA to reduce processing time and relax conservation and design controls (Head of Urban Design, Bristol City Council, pers. comm.). Similar trends were evident in the Birmingham area, where the rise in processing times in the mid-1980s was linked to attempts to develop design and conservation standards. The later reduction in time taken reflected the challenge to these policy developments resulting from appeal decisions against the LPA.

Appeals proved to be a significant indicator of local policy strength within the national planning and development context. During the 1980s, the appeals procedure came to form an increasingly important part of the planning decision-making process, and decisions resulting from such appeals had important implications for planning and conservation policies at the local level. In the Bristol conservation area, the number of appeals increased by more than half between the 1970s and 1980s, with ten and twenty-six in each decade respectively (table 3.1), in line with national trends (DoE 1991). This reflects the pressure put on Bristol's strong conservation controls in the 1980s. However, the strength of conservation arguments in central Bristol, with its 'priority town' status, allowed the LPA to sustain refusals for unacceptable major development and deflect the threat of appeal. Despite this, the LPA was less successful in sustaining refusals against minor development and changes of use in the conservation area, highlighting the nationally perceived limits to local conservation control policies in relation to business operations (Barrett 1996).

While the number of appeals in the Birmingham conservation area did not increase significantly between the 1970s and 1980s, there was an important change in the nature of the appeals from minor to major developments (table 3.2). As the Birmingham LPA sought to move away from its permissive stance of the 1970s and to tighten its conservation controls in the 1980s, the range of developments challenged increased, specifically to include more demolition, new building and major rebuilding. Of particular significance was the number of withdrawn appeals during this period, reflecting the use of the threat of appeals to pressurise LPAs to negotiate more rapidly. Also of note was the granting of appeals against refusal of major rebuilding and demolition. These appeal

Table 3.1 Appeals in Bristol, 1970s and 1980s

	1970s			*1980s*		
	Grant	*Refuse*	*Withdraw*	*Grant*	*Refuse*	*Withdraw*
ADD				1		2
COU				4	3	
DEM		1	1			2
MAJ		1				
MIN				1		
MIS				5		
NEW	2	1	2		1	3
REF	1					1
SIG		1		2	1	

Table 3.2 Appeals in Birmingham, 1970s and 1980s

	1970s			*1980s*		
	Grant	*Refuse*	*Withdraw*	*Grant*	*Refuse*	*Withdraw*
ADD	1					
COU		1	2	1	2	4
DEM			1	2		3
FAC				2		
INT				1		
MAJ				2	1	
NEW						3
SIG	5	5	6	2	1	1

Key to tables 3.1 and 3.2

ADD Extension (horizontal or vertical) to an existing building; building of new freestanding auxiliary structures
COU Change of use (between DoE use classes)
DEM Demolition of all or part of a building
FAC Alterations to building façade or shopfront
INT Internal alterations
MAJ Major rebuilding (demolition of all but a small part of a building and construction of a new structure)
MIN Minor alterations
MIS Miscellaneous non-fabric alterations
NEW New building (completely new structure on cleared site)
REF Refurbishment
SIG Illuminated and non-illuminated signs

decisions had important implications for the development of conservation policy. Following these decisions, the LPA was reluctant to refuse developments and go to appeal, owing principally to the introduction of the award of costs against the LPA.

Successful negotiation of central/local tensions: 16–20 Narrow Quay, Bristol

During the 1970s in Bristol, plans were drawn up for the refurbishment of the blighted City Docks. These plans actively sought to preserve and refurbish the eighteenth- and nineteenth-century pennant stone and 'Bristol Byzantine' (decorative polychrome brick) warehouses and dockside transit sheds in the area, and to build on the local Bristol dockside character through the promotion of new building styles that reflected these structures (Bristol City Council 1976). While this policy was successful in the late 1970s, commercial pressures by the mid-1980s had increasingly pushed styles of new building around the Dockside beyond the guidelines set out in planning briefs. The LPA acted to protect the essence of the 'docks vernacular' from dissolving too far into postmodern eclecticism, leading to a number of protracted negotiations on new developments around the Docks (Barrett 1996).

A notable example of the increased negotiation difficulties faced by local conservation officers was the proposed redevelopment of a sensitive site at 16–20 Narrow Quay, opposite the Watershed Arts Complex. In 1985, the LPA sought to 'enforce' the dockside character by refusing an application for a 'Georgian building in stone' for the site. The LPA also refused the two subsequent schemes, which although in a 'docks style' included too many storeys, had poor detailing and required the demolition of No 16, which was unacceptable to the LPA. In response to increasing developer interest in the site, a planning brief was produced for the site which proposed the retention of the eighteenth-century council-owned warehouse at 16, together with a new three-storey building for the cleared site of 18–20 (Bristol City Council 1987). The options for this were either an 'accurate reconstruction' of the Dutch-gabled elevation of the buildings demolished or a 'modern sympathetic infill' with rendered walls and stone dressings, although no example design was given. Crucially, the LPA was able to uphold its stance, and appeals against the refusal of demolition consent and new building consent were withdrawn, allowing the LPA to negotiate with new developers for their preferred scheme. This negotiation was still in progress at the end of 1989, when the downturn in the development market curtailed further consideration of the scheme.

In this case the combination of demolition controls, a well-argued case for the retention of 16 in terms of its historical importance and suitability for refurbishment and a clear design brief was critical in the LPA successfully upholding its stated conservation objectives against developer pressure. Generally, widespread building listing and use of detailed design briefs gave the Bristol area more success in curtailing the worst excesses of developer pressure, upholding earlier conservation policy gains and maintaining a distinct local character (Barrett 1996).

Unsuccessful negotiation of central/local tensions: Birmingham

The tensions between the application of local conservation policies and definitions of historic worth and the domination of national conservation criteria most clearly emerged during negotiations and appeals pertaining to major office development in the Birmingham area. In particular, the persistent use of façadism for the redevelopment of key listed Victorian buildings began to cause increasing conflict between LPA conservation initiatives and development interests in the 1980s (Birmingham City Council 1984; see also Barrett and Larkham 1994). In the office boom of the mid-1980s, the alliance of planning policies aimed at conserving the Victorian heritage and obtaining contextual solutions to new development and the desire to redevelop these buildings to meet modern office standards produced a pressure for façadist schemes. As the Draft Conservation Strategy noted, 'despite vigorous efforts, there is a continuing loss of many important buildings and interiors' (Birmingham City Council 1986).

The emerging tensions between local conservation objectives and the commercial demands of national developers came to a head during negotiations on an application for the redevelopment of 55–73 Colmore Row by Barclays Bank. The scheme involved the demolition of all but the façades and the banking hall of the listed Victorian Palazzo buildings owned by the bank along Colmore Row, Church Street, and part of Barwick Street at the heart of the conservation area (Barrett and Larkham 1994). The LPA demanded greater retention of the building structure, in line with earlier redevelopment schemes along Colmore Row. However, the developers did not wish to compromise and exerted considerable pressure on the LPA for a rapid settlement to the scheme through the submission of multiple applications. The original application submitted in 1985 was not determined within the statutory time period and was superseded by two parallel schemes in 1986, one of which was refused by the LPA, and one which was again not determined by the LPA within the statutory time period. The developer decided to take the non-determined applications to appeal to the DoE, following a breakdown in negotiations between the developer and the LPA. The schemes were granted on appeal, with the inspector indicating that they both satisfied conservation objectives, concluding that the façades alone were of real importance in satisfying conservation aims, and that the proposed buildings provided an economic re-use of the site. At the national level, the DoE regarded retention of the façade as an acceptable planning compromise, providing economic re-use of the site and satisfying conservation objectives in the case of mid-to-late nineteenth century structures listed for local importance only.

The decision to allow the façadist scheme at 55–73 Colmore Row was based on the general presumption in favour of development at the national legislative level, and the apparently lower importance accorded to Victorian architecture within the national context when tested at appeal. Nationally, Victorian and Edwardian architecture has not (until relatively recently) generally been conserved to the same degree as have buildings of the eighteenth century. In the Birmingham area, much of the fabric post-dates 1850 and is not highly regarded as preservable at the national level, despite its crucial local importance in establishing the character of the Colmore Row Conservation Area. In this case, the minimum of public conservation demand was satisfied, at the expense of a loss of meaning for many key buildings in the Colmore Row area with a significance to the character of the area in socio-historical terms that went beyond their architectural merit. The developer's ability to appeal to the national context within the local development control system highlights the way in which national values overrode local concerns in conservation area control. Locally, the 55–73 Colmore Row decision left conservation policy in the area in a weakened state, with regard to control over demolition and moves to obtain greater preservation of the Victorian fabric (Head of Conservation, Birmingham City Council, pers. comm.). In the face of continued undervaluation of the area's fabric, and a continued pre-eminence of economic arguments over conservation concerns in the core, the LPA remained unable to stem the erosion of the Victorian fabric, concentrating instead on using its limited ability to control demolition to influence stylistic considerations.

Market forces: enhancement and conservation gain

In the 1970s and into the early 1980s, conservation area enhancement strategies in both areas principally took the form of positive action in terms of landscaping improvements. This was a pump-priming exercise which, it was hoped, would stimulate private-sector-led improvements to buildings. With its injection of grant funds in the late 1970s, with which to initiate landscaping improvements, the Bristol area led the way in this form of development, in advance of efforts in the Birmingham area. However, in the mid-to-late 1980s, LPA policy in both areas shifted towards the management of private-sector-initiated building 'enhancements' (Barrett 1996). Both LPAs sought to shift the responsibility for high-quality refurbishments, both to the interior and exterior of buildings, to the private sector, given the increasing commercial interest in heritage. With a reduction in grant aid from both central and local government, the use of conservation gain became increasingly important in obtaining refurbishment. This mirrored the wider adoption of planning gain as a strategy used by planners to obtain added social and environmental benefits from developers in the 1980s.

Examination of LPA efforts to control and direct minor change and private enhancement activity in the conservation areas is particularly useful in exploring the limits to LPA conservation controls in relation to business demands. It is clear from examination of negotiation on minor changes in both study areas that the degree of relative power between conservation control and efficient business operation remained blurred. In the 1980s, both LPAs encountered difficulties in applying strong control policies over minor change deemed unsuitable and in obtaining conservation gains from private applicants in return for planning permission, consequently limiting enhancement activity. The ability of LPAs to develop strong policies to counter erosion of character through minor change and enhance conservation areas was constrained by the nature of planning legislation and the attitudes of central government to development. Only in the case of listed buildings are extra controls offered, tying success in controlling change to the national worth placed on the fabric of a conservation area. Planners were, therefore, dependent on the general planning controls available to regulate minor changes; but many of these were 'permitted development' and did not require specific planning permission. This increased negotiation difficulties for LPAs, which intensified throughout the 1980s as central government sought to increase permitted development linked to deregulation in the planning system.[3] The lack of wider controls meant that, in both conservation areas, much minor change remained unmonitored and outside LPA control. In particular, problems arose in the 1980s with the positioning of 'heritage clutter', such as non-illuminated signs, on buildings, much of which was classed as permitted development.

Conservation gain and enhancement in Bristol: the removal of forecourt parking, Queen Square

One of the key problems identified within the City and Queen Square Conservation Area as a threat to area character was the incursion of cars and the demand for parking. Throughout the 1970s and 1980s, the Bristol LPA sought to control forecourt parking in Queen Square, to complete environmental improvements in the area: 'the impact of parked vehicles within historic areas should be minimised . . . the intrusion of vehicles into front gardens or forecourt areas will be opposed' (Bristol City Council 1984: 19). The wider enhancement of Queen Square initiated in the early 1990s constituted a significant improvement to the conservation area. However, this was not achieved without a considerable battle on the part of the LPA, and it was only when an enhancement strategy was incorporated into the statutory City Centre Local Plan that significant advances were made (Bristol City Council 1990). Prior to this, change was more incremental and piecemeal, with conservation gain demands tied to office development used to obtain improvements. During the mid-1980s, the pressure for increased office space in the Square was used as a lever with which to strike deals for the removal of parking and the restoration of forecourts. However, limits to the application of this strategy became evident. In 1982, attempts to obtain wider conservation gains beyond the change applied for were challenged, when the refusal of an application relating to 46 Queen Square that did not meet the wider conservation gain requirements of the LPA was taken to appeal. The minor change was granted on appeal, as the LPA was seen to be demanding too much from the applicant; it was accepted that there was no right to refuse permission on the grounds of non-compliance with changes demanded beyond those for which application had been made. This condition was again challenged in 1986, at the height of the development boom, when negotiation on a new application failed to secure parking removal with the refurbishment of 46 Queen Square, as it was claimed that refurbishment already provided a conservation gain. In 1984, the refusal of an application to remove the parking condition imposed during negotiation for a development at 49–51 Queen Square was granted on appeal. Here it was felt that the developers had provided sufficient conservation gain in the refurbishment of the building, in line with the LPA policy. Clearly, Bristol's conservation strategies encountered problems when pitched against commercial imperatives, such as the demand for parking.

Conservation gain and enhancement in Birmingham: shopfront control in New Street and Corporation Street

In the 1980s, the Birmingham LPA sought to develop control and enhancement policies in the primary retail streets containing the nineteenth-century fabric, included within the conservation area in 1984. However, increasing control of applications for shopfronts and signs proved controversial, reflecting an important intersection between business and conservation demands at the micro-scale. In the late 1980s, appeal losses on major developments made the LPA reluctant to use 'tied' permission or the possibility of withholding or delaying permission as a 'stick' with which to obtain further conservation gains beyond those changes applied for. Therefore, the LPA sought to encourage the removal of the existing legacy of poor shopfronts and signs and foster a climate of good design through the offer of grant aid, given the limits to the improvements obtainable through strong control and enforcement action. Grant aid was used as a 'carrot' to persuade applicants to improve the quality of designs submitted, or to provide enhancements beyond the changes for which permission was sought. In a number of instances, grant aid was offered to achieve a design solution in line with LPA policy, and offset the reluctance of applicants stemming from the perceived cost of refurbishment schemes. In the conversion and refurbishment of a former cinema in New Street, grant aid was used to obtain the design stipulation of hand-painted signs and wood frames. Grant aid was also used to expose terracotta and add pilasters to shopfront schemes on important Arts and Crafts buildings in New Street.

However, the wider development of these retail area enhancement policies proved problematic in the prevailing aggressive business climate of the 1980s and given the lack of funds with which to encourage refurbishment. As an internal report to the CAAC into the attempts to encourage the refurbishment and re-use of the Victorian buildings in the area noted, 'without substantial grant aid and the support in appropriate appeal cases of the Secretary of State, such a trend is unlikely to develop' (Birmingham CAAC 1984). Consequently, improvement remained dependent on co-operation from applicants and private landowners and strong pre- and post-submission negotiation by the LPA. Efforts to develop integrated shopfront improvement schemes on blocks of Victorian buildings met with limited success. Where the City Council owned the freehold of buildings in the retail streets some success was achieved in developing integrated refurbishment schemes. However, in blocks in private ownership, many tenants were unwilling to develop schemes unless applying for change themselves. In Corporation Street, the LPA pressed for the uptake of grant aid by refusing a shopfront application when the applicant declined the offer of grant aid – a risky strategy given the danger of appeal on these refusal grounds. Compelling applicants to engage in enhancement strategies proved difficult given the lack of support from the private landowners of these Victorian commercial buildings towards their enhancement. The actions of the private landowners of 2–6 Corporation Street effectively stopped the wider development of refurbishment schemes in lower Corporation Street, as they did not wish to improve their building whilst waiting for the opportunity to redevelop. The application of sign and shopfront policies here exposed both the limits to the application of conservation and design controls in the face of commercial pressure, and the wider limits to negotiation and control over minor change in the planning system. In particular, it highlights the limited effectiveness of the conservation area in adding to the LPA's power to control minor change and the power of private landowners in commercial areas to limit the impact of conservation policy at this time.

Conclusions

This chapter argues that conservation under the Conservatives has suffered mixed fortunes, although there was, clearly, some progress. The key tenets of the Thatcher/New Right agenda are clearly displayed: throughout the period there has been increasing centralisation of control through 'guidance' documents, new quangos, a new government department, and high-profile personal interventions by ministers. History and heritage became important political considerations, although Thatcher's own interpretation of history was very selective. As Raban observed,

> her own break with the past has been radical to the point of being revolutionary, yet she continually employs 'history' as the great licensing authority, to validate every departure from historical practice. Her notions of what actually happened in history are often eccentric, sometimes downright ignorant. At fonder moments, she substitutes the phrase 'our heritage' for 'history' and 'heritage' expresses her meaning more clearly. For a heritage is someting we have possession of after the death of the original owners, and we are free to use it as we choose.
>
> (Raban 1989: 23–4)

The amount of funding for conservation has declined in real terms, offset to some extent by inventiveness in using other budgets for conservation purposes, but exacerbated by this tightening of central control. Market forces have become significant, through compulsory competitive tendering, privatisation, and the increasing commodification of heritage. Nevertheless, this is also a period when the number of listed buildings (designated centrally) and conservation areas (designated locally) have risen dramatically, and continue to grow; and when the activities of numerous pressure groups and the popularity of conservation/heritage tourism suggests that its popularity amongst the general public is also at a high level. However, this 'public' is often an educated élite in vociferous and well-informed pressure groups, which has not let the continuing Thatcherite agenda go unchallenged, in court if need be. This group was successful in securing at least one U-turn, in revising the control of minor developments under the General Permitted Development Order in 1995.

The entry of the New Right into the delicate balance of central/local élite views served to downgrade conservation generally, to centralise planning and to reduce participation. This occurred, for example, with the increase in channelling funding through quangos, including English Heritage (there were 5,521 such quangos by 1994, despite the Thatcherite rhetoric of 'rolling back the frontiers of the state': Hewison 1995: 229–30) and through the pressure of the eight-week period for processing applications. Thus many aspects of control were moved from local élites to a central élite, although this was continually challenged at the local level through planning negotiations, appeals etc., as the examples here show. Locally, conservation remains an élite activity. Whilst its influence has grown within planning, it has not become particularly democratised: planning remains hierarchical, participation has been downgraded through the 1980s, and conservation remains bound up in the debate over national and local élite heritages. There has not been the will, time or resources to effect fundamental change. The New Right simply sought to consolidate this position through limiting control at the local level, and moved the balance back towards the centre from the local: although with mixed results, as the examples demonstrate.

Similarly, although the rise of the 'plan-led' planning system in the 1990s moved away from the *laissez-faire* of the Thatcherite 1980s, it has been strongly argued that much conservation activity should – and does – occur outside the potentially rigid plan framework. Conservation does not fit easily even within the more conciliatory, less developer-friendly, system of the Major era.

As much conservation also stresses the distinctiveness of localities and communities, it is at this scale (rather than the national scale) that successes and failures can be tested. Even here there are clear conflicts between national policies and interpretations, and local distinctiveness and desires. Evidence from the two case-study areas suggests that conservation areas continued to function as areas of special control within an increasingly deregulated system. However, where conservation concerns intersected with business demands, the confused messages from central government created an arena of contestation as the balance between business and conservation was reformulated at the local level. The abilities of individual LPAs to formulate their own policies and approaches, or to support these through funding (the 'implementation perspective'), and their relationship to changing national approaches remains under-researched. Nevertheless, unlike other policy areas such as urban regeneration, conservation suffered relatively little from a Thatcherite top-down directed approach: its success has always been recognised to be its local strength.

Probably no-one on the New Right has had such a well-publicised clash with conservation than Theresa Gorman, MP for Billericay. She had carried out £300,000 worth of alterations to her Grade II listed farmhouse, but had forgotten to apply for planning permission and listed building consent: saying 'as an MP you are a very busy person' (quoted in Ezard 1995: 22). Prosecution was agreed by the LPA because of the scale and nature of the alterations, and because work continued after a warning to stop. She disagreed with the LPA's assessment of the changes, but was ordered to carry out amendments when six enforcement notices were upheld, six more being quashed subject to further alterations being made (*The Guardian*, 1 December 1995: 6). As a result of this still-unresolved battle, Gorman has campaigned against all conservation planning control: commenting upon the 1996 Heritage Green Paper that:

> it's really an eye-opener because it takes for granted that the role of the Government is to interfere and protect listed buildings . . . the tone of the document is extremely authoritarian, without hardly a note of the fact that what we are talking about is

> privately-owned property from which the state has removed the rights of the owner.
>
> (quoted in Hirst 1996: 13)

These views are extreme, much more so than those even of the former Secretary of State Nicholas Ridley. They not only reflect a grass-roots conservative pro-enterprise culture, but propose dismantling the very mechanism of protection and control which appears to be supported by so many middle-class/élite of the local amenity movement, often misrepresented as NIMBYs (and many of whom have traditionally been Conservative voters). Although Michael Heseltine as Secretary of State made some pro-developer decisions, including introduction of the Certificate of Immunity from Listing, he was, for the most part, much more representative of these majority traditional conservative views.

Whilst Gorman's views are extreme, spurred perhaps by personal experience, in other policy fields she is more central to the New Right philosophies. She clashed fundamentally with others of her own party in the mid-1990s, including Heritage Secretary Virginia Bottomley and Environment Secretary John Gummer. Her extremity shows the unusual nature of conservation planning during the Thatcher/New Right decade of the 1980s: that, in this aspect, planning did not wholly fail. Local diversity and approaches remained, despite clashes with the centre. The Major years saw continuation of the New Right centralisation (e.g. the new DNH) but, crucially, the more recent acceptance by ministers that local planning is necessary. Conservation survived the New Right because of its broad local basis of support.

Notes

1 Cullingworth and Nadin (1994: 86) note that, at the same time that government publications exhorted deregulation, 'Perversely, the number of appeals received rose during this decade, reaching a peak of 32,281 in 1989/90, since when there has been a significant fall. The increase in appeals had the effect of slowing down the appeals process.'

2 It should be noted that variations are also linked to fluctuations in the amount of major development activity in an area, the strength of policy development and the extent of consultation and negotiation exercises. Differences in type of change also account for differences between the two areas, with the longer average time taken in Bristol partly due to the greater volume of development activity. The numerical dominance of sign applications in Birmingham partly accounts for the lower processing times. See Larkham (1991) for a wider discussion of delays in processing times.

3 A revision of conservation-related permitted development rights did occur in the 1995 General Permitted Development Order, giving additional power to the LPA at the expense of the Secretary of State. This was seen as a surprising U-turn by ministers and DoE alike (see commentary in Larkham and Chapman 1996: 15–16).

4

PLANNING FOR HOUSING: REGULATION ENTRENCHED?

Glen Bramley and Christine Lambert

Introduction

Housing is the most important urban land use and accounts for the greater amount of land subject to urban development in most areas and periods (Shepherd and Bibby 1996), even in a mature economy with slow demographic growth like Britain. The evolution of the planning system historically was closely bound up with the evolution of housing policy and provision. For these reasons the way the planning system deals with new housing development provides an important test for propositions about the impact of New Right ideas and policies on planning. Yet much of the literature and debate about changes in planning in the 1980s focuses on other areas, such as urban regeneration and economic development, perhaps because these had a higher profile and provided examples of radical change. In this chapter we argue that housing remains a key arena for the practical application of planning, and that the picture emerging is one which seriously calls into question notions of radical change or the side-lining of planning. This is not to say that there have not been significant changes in certain respects, and we illustrate these through local examples. However, at the same time we point to evidence of both continuity and, in some respects, reassertion of the core regulatory function of planning, despite the rhetoric of the New Right.

We start by asking what a 'New Right' philosophy might look like and what its general programme is? A number of strands can be identified, some rooted more in economic ideas and some more in the political realm. What does this set of ideas imply for planning, and how would we expect this to be manifested in planning policies with particular relevance to new housing? In discussing these ideas in general terms it quickly becomes clear that there are some contradictions within the New Right approach, contradictions that are perhaps particularly salient in the case of housing.

We then go on to review what actually happened to planning policies and practices relating to new housing, generally and at national level, over the period 1979 to 1997. How far was this picture actually consistent with the prescriptions of the New Right, and if not, why not? What does this tell us about the adequacy of New Right thinking or about the correctness of this as a characterisation of policies? This review suggests a significant change in emphasis between the earlier and later parts of the period, although the change of emphasis probably predated the change of leadership from Thatcher to Major and arguably arose out of the limitations and contradictions apparent within the original thrust of New Right policies.

The chapter then goes on to examine what happened at local level in certain case-study areas. We look in particular at two areas in the south of England subject to major housing development pressures, Swindon and the Bristol subregion. This discussion is also informed by wider statistical evidence on patterns of planning restraint and its impact on housing, and by other literature covering other localities, again with a general emphasis on areas of high housing growth and pressure. The case studies highlight the importance of local context and historical continuities, as well as the critical role of land ownership, infrastructure financing and the economic instability of the housing market.

New Right thinking

Economic ideas

Two major strands of New Right thinking are identified here, one growing out of economics and the other more focused on politics. The first of these refers to the 'public choice' school, which essentially applies neoclassical economic assumptions and analytical methods to political and organisational behaviour. Thus this approach adopts an individualistic theory and assumes that individuals take rational decisions that maximise their individual interests subject to limitations of knowledge etc. These assumptions are inherently pessimistic about the possibility of planning acting in its traditionally assigned role of acting in the public interest.

From the 'economics of politics' school (Downs 1957; Buchanan and Tullock 1962; Buchanan 1978) are derived a number of propositions about the biases and shortcomings of representative democracy, whether at national or local level. For example, majority-based decision-making may reflect the interest of the median voter but will not necessarily produce efficient outcomes. Overall, this school adopts a critical stance towards democratic decision-making, which underpins a general presumption towards limiting the scope for decisions to be made in this collective way, for example by limiting the role of local government, or by privatisation and marketisation of state activities.

A second major strand is the 'economics of bureaucracy' school (Niskanen

(1971), Downs (1967), Jackson (1982) all provide more critical reviews). The main thrust of this school is to argue that bureaucrats act in their own private interest, which may take the form of empire-building and budget-maximisation, and that they use their control over information in the relationship with their political masters to achieve this end. This is expected to lead to a systematic overproduction of the bureaucratic goods and to productive inefficiency in their activities. Recommended solutions include privatisation, the introduction of external or internal competition, and better provision of information and mechanisms of accountability.

The critique of professionalism is an added element here, which is clearly important in planning. This aspect of New Right thinking was a characteristic and new feature in the Thatcher era, but interestingly connects with some new left thinking of the same era. Distrust of professional expertise may be a broader phenomenon of this era, allied to the decline of deference and disillusionment with scientific expertise and the notion of progress, and as such part of the overall postmodernism phenomenon.

State intervention has traditionally been justified by appealing to a welfare-economic paradigm of market failure. What the New Right brought to bear was a counter-attack in the form of a theory of 'state failure' (Wolff 1988; Le Grand 1991). Thus, for example, regulatory interventions like planning could systematically fail because of such factors as 'regulator capture', information problems, restrictions on entry to markets, producer dominance, and price distortions (Bramley *et al.* (1995: ch. 10) review these arguments in relation to planning for housing).

Planning's justification can be firmly located in the market failure paradigm, referring to endemic failures associated with local public goods, the land market and urban/environmental externalities in particular (Walker 1981; Harrison 1977). The economic logic of this argument is one which the public choice school cannot wholly deny. One can interpret the differential impact of the New Right on different sectors of the welfare state in this way: for example, the vulnerability of housing to privatisation because of its weaker market failure case, compared with health. Land-use planning has stronger claims still on market failure grounds than state housing provision. However, the information requirements of successful planning intervention are formidable, and this weakness echoes arguments of Hayek dating back to the 1940s that the project of a planned economy was inherently unachievable in information terms. Hayek's ideas became influential again in the 1980s after lying dormant in the earlier period. The corollary of this is a much revived and widely shared confidence in the virtues of markets as mechanisms for processing information and meeting consumer preferences. This has been underlined by the spectacular failure of planned economies and their new-found enthusiasm for markets, as well as by the growing sophistication and diversity of consumer preferences and needs (postmodernism again).

The New Right is also associated with macro-economic prescriptions for reduced public spending, borrowing and taxation, and 'supply side' reforms to labour, finance and other markets. Some of these reforms have indirect implications for housing markets, for example the deregulation of mortgage lending. However, the most obvious effect of the macro-economics of the New Right has been strong downward pressure on public expenditure with a particular emphasis on public investment and borrowing.

Last but not least among the economic strands, the New Right can be associated with a different attitude to economic and social inequality, and a markedly different performance in distributional outcomes (Hills 1996). The conventional justification is that this is a price worth paying for more economic dynamism and competitiveness, as well as by referring to social theories which emphasise the evils of dependency or cultures of poverty.

Political ideas

Turning from economic to political ideas, it is generally accepted that New Right ideology is composed of two main strands: a neo-liberal strand that emphasises individual freedom within the context of a free market system, and a neo-conservative strand emphasising the importance of social order, discipline and authority (what Hall and Jacques (1983) call a 'social authoritarian' and Gamble (1994) refers to as the 'free economy and the strong state'). The logic is freer markets and limited government, but a readiness to use the powers of the state to confront those groups who resist economic reforms. The assault on local government in the 1980s can be seen as encompassing both aspects.

It is also appreciated that there is some tension (or even contradiction) between these elements of New Right thinking, a contradiction illustrated in the planning area by the Green Belt controversy in the 1980s. In relation to the conservative strand, there is often a special place given to the British landscape as symbolic of certain national traditions, in that order and tradition are often associated with rurality. This element could however be characterised as belonging to an earlier 'old right' tradition of conservatism upheld in particular by landed interests. Others might argue that New Right housing policies gave rise to the growth of a strong home-owner lobby with direct interests in protecting its property interest. Again, within the conservative strand there is emphasis on property as one of the foundations of authority and order, while within the libertarian strand it is seen as a foundation of liberty. This example would seem to be one where there is an inherent contradiction within the New Right, because upholding owner occupiers' rights creates a powerful vested interest which may resist economic change.

One argument that may be put forward to overcome this problem of contradiction is that it is entirely appropriate to intervene selectively to accommodate different circumstances. So in special and valued areas, where landscapes and townscapes are of high quality, strict controls are appropriate; in areas where the market needs pump-priming before it can work, intervention is justified; and elsewhere market-led development is appropriate. This disaggregation of areas and introduction of different planning regimes was characteristic of planning in the 1980s (Thornley 1991; Brindley *et al.* 1996).

On the other hand perhaps we should not expect governments to implement coherent packages of policies consistent with particular philosophical positions. Both theory (including 'public choice' theory) and experience teach that politics is often rather pragmatic, instrumental and liable to proceed by 'disjointed incremental' steps as certain trade-offs are made to achieve change. The resulting policies, not always ideologically pure, are consistent with a view that Thatcherism was as much as anything a political project, aimed at re-establishing Conservative political leadership. It involved statecraft as well as ideology (Gamble 1994), guided by a set of intellectual ideas, but also developing a programme of policies that responded to the interests and concerns of voters and interest groups necessary to continued electoral success (the sale of council houses is a good example). The 1980s can perhaps be better interpreted as moving strategically towards certain New Right principles while making certain tactical compromises, effectively 'rolling back the state' rather than dismantling it.

Implications for planning for housing

It is straightforward to derive a number of basic propositions about the kinds of planning policies and approaches to housing development which are implied by the beliefs of the New Right.

A *reduced role for the state in providing and financing housing* is the first obvious implication of this package of beliefs. Such a change offers opportunities to cut public expenditure, borrowing (in particular), and taxation. It exploits the emerging weaknesses and unpopularity of public housing which appeared to lend support to the New Right critique of public bureaucracy and the allied professions, particularly in relation to the consumers' lack of choice and control. Reducing public housing directly reduces the scale of the public bureaucracy and reduces the scale and scope of that deeply suspect institution, local government. It increases the demand for privately provided housing, particularly owner occupation, and extends the scope of the market.

As we show below, this prescription has been followed over the period of study, with direct public housing investment reduced to a fraction of the level characteristic of the period 1945–75. The main implication of this change for planning was substantially to remove the option of 'public-investment led planning' (Brindley *et al.* 1996) where housing might have been a major element. Traditionally, major housing developments were often led and dominated by public provision, enabling a 'command and control' mode of planning, or what is sometimes referred to as 'positive planning'. The loss of this option means much more exclusive reliance on the responsive mode of planning, through development control on the basis of structure and local plans, although in some cases leverage or partnership modes might be employed (particularly in difficult regeneration projects).

The second and complementary prescription is that the state should *facilitate the development of private-sector housing provision*. This is motivated in New Right thinking by the general presumption in favour of extending markets, increasing competition and responsiveness on the supply side and consumer choice and control on the demand side. There are additional and perhaps more covert political motivations, which are less respectable in philosophical terms but may have been important in practice: new home-owners may change their voting behaviour, and particular producer interests are favoured.

We review below the means adopted to achieve this aim, which have been broadly successful, so that owner occupation has come to be the dominant housing tenure and the dominant form of new housing development.

The third and most critical element of any New Right programme for planning must be to *reduce excessive and unnecessary planning regulation* affecting new housing development, including possible delays and costs imposed by procedures and by negotiations with planning authorities. The whole force of New Right thinking is to be critical of bureaucratic or professionally based regulation of the market, perhaps particularly where this is allied to local democratic control through the suspect institution of local government.

A whole raft of measures could be seen as falling under this general rubric. These fall into three broad categories: those which have actually been implemented, those which have seen partial implementation, and those which, while they may have been discussed and even attempted, have not in practice been implemented over the period under study. Overall, as we show below, the picture is very mixed in terms of the overall extent of reduction in planning-based regulation. Particularly interesting, and potentially contradictory, is the 1990s move to a 'plan-led system'.

A further element in the New Right agenda is undoubtedly the aim of *removing or reducing the role of local government* in planning decision-making and implementation. A number of the philosophical and theoretical strands feed into this negative view of local government: critical views of the democratic process; vulnerability to capture by special interests; the bureaucratic character of organisations, and/or professional dominance; perceived incompetence and inadequacy to the challenge of planning, associated with élitist views of 'councillor calibre' and officer expertise; fear of political control by the ideological left or attempts to pursue inappropriate redistributive policies.

We argue below that this thrust of policy has been less evident in practice in the case of planning for housing than in some other areas. There have been some examples of stronger central guidance and interference in decisions, but in general the key focus of planning decision-making remains local.

While accepting the existence of public goods and externality problems, the New Right may be expected *to favour economic mechanisms for resolving environmental externality problems* rather than the traditional emphasis on regulation. This preference over means reflects the economistic emphasis in New Right thinking and the general preference neoclassical economists display for price mechanisms on efficiency grounds. Broadly the polluter should pay, and decide whether the activity is worth it, rather than some bureaucrat decide. This issue is more relevant to other sectors and land uses than housing, and raises issues for other public agencies and professions in addition to planning and planners. Overall, despite rhetoric, the extent to which taxing and charging mechanisms are used to deter adverse environmental effects remains limited in Britain: for example, increases in fuel duties to deter energy use with its adverse global and local effects, are quite modest. Nevertheless, part of the role of planning agreements is in effect to act as a (negotiative) pricing mechanism for either preventing or making good adverse environmental consequences.

The New Right enthusiastically endorses privatisation in almost all circumstances, and consequently measures *to hand over significant parts of the planning function to the private sector* (Brindley *et al.*'s 1996 'private management planning') would be expected to comprise part of the programme. Few corners of the public sector have escaped the privatisers' attentions, and great ingenuity has been applied to effecting the privatisation of some fairly implausible candidates, sometimes at great cost. Yet, as we shall see in the case of planning, there has been relatively little shift in this direction.

The extent of change in practice

We now review the extent of change observed in practice over the system as a whole, in terms of these general propositions about what we would expect from a New Right agenda for planning for housing.

There has been a major *cutback in direct public housing investment*. This change was effected by reducing and eventually eliminating the ability of local authorities to build new housing. It was further reinforced by the running down of the New Towns programme and the rundown of comprehensive housing redevelopment. Most of this change was effected in the earlier (Thatcher) part of the period under study, and indeed it was initiated before that in the last years of the 1970s Labour government. In the early 1990s there were some offsetting increases in publicly funded housing provision by housing associations (quasi-private sector), and large-scale increases in Housing Benefit (rent subsidy), much of which went to the private rented sector. There were also interesting developments in the use of the planning system to lever in social housing provision (PPG 3, DoE 1992), discussed further in the local context below. But overall these recent developments have not offset the major decline of public housing investment, as shown in table 4.1.

The public sector has thus largely withdrawn from the direct land-development role: for example, in the former New Towns or through such

Table 4.1 Housing public-expenditure time trends

Year (financial)	*Housing public expenditure 94–5 prices* £bn	*Total public expenditure 94–5 prices* £bn	*Housing share of public expenditure* %	*New housing completions GB* *Public* '000	*Social* '000	*Public share* %	*Social share* %
1971	7.8	141.1	5.5	150	160	45.5	48.5
1972	8.9	148.4	6.0	118	125	37.1	39.3
1973	13.9	160.0	8.7	100	109	32.6	35.5
1974	19.5	177.9	11.0	120	131	38.0	41.5
1975	16.2	178.0	9.1	140	157	47.0	52.7
1976	15.3	174.4	8.8	140	159	45.5	51.5
1977	13.2	163.6	8.1	130	154	42.6	50.5
1978	13.6	176.9	7.7	102	124	38.5	46.8
1979	10.2	172.9	5.9	84	104	33.3	41.3
1980	11.8	229.2	5.1	86	107	36.5	45.5
1981	8.1	232.1	3.5	66	85	32.8	42.5
1982	6.9	238.5	2.9	37	50	21.2	28.6
1983	7.6	241.3	3.1	35	51	17.6	25.7
1984	7.3	246.7	3.0	34	51	16.2	24.1
1985	6.3	245.8	2.6	27	40	13.8	20.5
1986	6.0	247.7	2.4	23	35	11.0	17.2
1987	5.9	248.0	2.4	20	33	9.3	15.1
1988	4.3	240.0	1.8	20	33	8.5	14.0
1989	6.4	251.5	2.5	18	32	8.3	14.9
1990	5.6	252.9	2.2	17	34	8.5	17.2
1991	6.2	257.8	2.4	10	30	5.6	16.3
1992	6.5	273.2	2.4	5	30	2.8	17.5
1993	5.3	282.5	1.9	3	37	1.4	20.7
1994	5.1	288.0	1.8	2	36	1.0	19.9
1995	4.7	294.5	1.6	2	40	1.1	21.4

Sources: Wilcox 1996: *Housing Policy Review 1996/97* Tables 14b and 18h; Goodchild and Karn 1997: Charts 1 and 2

mechanisms as the short-lived Community Land Scheme. Urban Development Corporations provide an exception to this generalisation. It is also true that some local authorities have continued to act as significant land developers by using existing land banks or land released from other uses, although in the long run this supply will dry up in most areas (Farthing *et al.* 1996 – see also case study of Swindon below). Measures to encourage authorities to dispose of such land (e.g. land registers, incentives relating to capital receipts) may paradoxically have increased the land development role in the short term. While falling short of the full public housing development role, this activity still gives greater control than purely responsive planning, because of the 'landowner control' element, but this may be compromised by financial pressures to maximise returns.

Government policies have clearly served to promote private sector housing provision, particularly the extension of owner occupation. Some of the mechanisms involved here are fiscal in character: various tax reliefs for homeowners (MIRAS, capital gains and imputed rental income exemptions) and potentially for private landlords, and subsidies for low-cost homeownership schemes. These do not fully satisfy the New Right's goals, because they worsen the public deficit, but may be justified on transitional grounds; it is interesting to note that in the 1990s one of these reliefs (MIRAS) is being scaled down. A major mechanism for increasing owner occupation has been the Right to Buy scheme, although this also has dubious effects on the public purse in the long term and does little to promote new private provision. Other measures to promote owner occupation, such as mortgage-market deregulation, are fiscally neutral and relate to wider New Right programmes of deregulation, but arguably create some problems of instability in the housing market and the macroeconomy (Maclennan 1994).

The implications of this programme for planning are broadly to reinforce the emphasis on responsive planning (or 'trend planning' as defined in Brindley *et al.* 1996) in relation to large-scale speculative private housing development. There is a presumption that the planning system, insofar as it remains in place, will be used to facilitate rather than block this type of development. Thus the moves to streamline planning procedures require more attention to land availability and market demand, and to incorporate the housebuilding industry in the process may be seen as in part a product of this general policy (Circulars 9/80, 22/80, 23/81). Similar comments would apply to the emphasis on meeting requirements for projected numbers of new households through regional planning guidance (DoE 1992) and central intervention in structure plans or major appeals.

Debates concerning planning for housing in the 1980s, particularly the issue of housing land supply, have been extensively reviewed elsewhere (Rydin 1986; Monk *et al.* 1991; Bramley *et al.* 1995). From early on the new Conservative government exhorted planning authorities to increase the supply of marketable land for housebuilders and to intervene less in matters of detailed design (Circulars 9/80 and 22/80). Exhortations to release more housing land have a long history in planning and were also a feature of the 1960s and 1970s. However, the emphasis in the 1980s was that planning should pay more attention to 'market considerations', as well as overall considerations of need and supply in determining the amount and location of new housing development. The main policy innovations were: joint housing land studies giving housebuilders' representatives more 'voice' in policy-making; the requirement that planning make a 'five-year supply' of housing land available; and that market criteria be used in determining 'availability'. In Circular 22/84 the DoE stated a 'special presumption' in favour of releasing land for housing development.

Nevertheless, guidance continued to require the balancing of housing demand with the need for environmental protection, and strict controls in Green Belts and other designated areas, together with policies to preserve open countryside, were maintained. By the early 1990s this was reinforced by important changes to the previous advice. The revised PPG 3 on housing (DoE 1992) withdrew the 'special presumption' in favour of land release for new housing, and signalled the first shift of emphasis to the re-use of urban land. Further policy guidance in PPG 13 (DoE 1994) placed further emphasis on urban redevelopment to reduce the need for travel, and also appeared to rule out the possibility of free-standing new settlements, unless proposals were able to demonstrate a high degree of self-containment. This advice can be seen as restating the long established principle of urban containment on which the planning system has been based (Farthing 1996).

The idea of *reducing planning regulation* must be seen as a central part of the New Right agenda. Here we find a much more mixed picture, with some examples of measures of deregulation being implemented, other partial or ambiguous examples, and other cases of no significant change. In the first category come measures such as the relaxation of the General Development Order (GDO) and the establishment of development control performance targets and regimes of indicators. The former seeks to roll back the boundary of regulatory intervention on the margin of detailed and small-scale conversion activity. The latter attacks bureaucratic inertia and delay through an information-based approach.

In an intermediate category of partial implementation come attempts at dropping some of the tiers of planning, moves to reduce or eliminate planning control in selected areas, and moves towards a more legalistic zoning system which reduces local authority discretion. The abolition of the Greater London Council and the metropolitan counties in 1986 removed one tier of planning authority in these areas and partially merged structure and local plan preparation at the district/borough level. Similar changes may occur in areas affected by local government reorganisation which created unitary authorities from 1996 onwards; this has been the normal pattern in Scotland and Wales, although here joint arrangements for structure planning remain in some areas. Proposals were mooted in a White Paper on *The Future of Development Plans* (DoE 1989a) to abolish structure plans; these have not been implemented, although there has been some simplification of procedures.

The radical concept of 'planning-free zones' has been instituted through two initiatives, Enterprise Zones (EZ) and Simplified Planning Zones (SPZ), the former being targeted on derelict industrial or commercial areas and the latter being very experimental. The experience here (Allmendinger 1997) emphasises the difficulties of wholly dispensing with planning, the range of interests (including land and property, residents, etc.) that rely on planning, and the implausibility of scenarios of general abolition of planning. These initiatives can be seen as marginal, tokenistic and watered-down versions of the original concept.

The concept of moving towards a zoning type of system is probably the most significant of the moves which have been at least partially implemented. Zoning is characteristic of market-dominated systems like the USA, and involves the designation of specific areas of land for particular uses subject to given development parameters (e.g. plot sizes and ratios, building lines). Its essential feature is that on zoned land there is effectively a right to develop within the defined parameters; the local planning authority does not have the ability to refuse permission on substantive or subjective grounds, only on grounds of nonconformity with the zoning. A zoning system substantially changes the relationship between planners and developers, compared with the situation under a discretionary development control system like that used in Britain (Grant 1991, 1992), particularly where there is no operative local plan containing allocated land. In England in the 1980s comprehensive and up-to-date local plans were the exception rather than the rule, but at the beginning of the 1990s the government set the target of achieving full coverage by 1996. This was part of the move to a 'plan-led system' which also encapsulated changes embodied in the 1991 Planning and Compensation Act giving priority in development-control decision-making to the provisions of the plan. We would see this change as being potentially important for housing, the largest single consumer of land subject to development and relatively amenable to zoning-type approaches.

The move to a plan-led system is in one sense surprising, because it seems to represent a substantial increase in the role and strength of the planning system, and as such to fly in the face of the New Right prescription of rolling back regulatory interventions like planning. There are two main views of the plan-led system. One view would be that this is a significant change, and that it does make British planning more like a zoning system (although still not the same as this in its pure form). From this view, the increase in bureaucracy and political decision-making at the plan-making stage is more than offset by the reduction in uncertainty, delay and negotiation at the development control stage; the net effect on the supply side is positive. However, implicit in this view is the argument that some form of zoning is optimal for efficient development in market economies, because of its benefits in reducing uncertainty and curbing adverse externalities. This argument in turn is a significant challenge to pure or simplistic New Right views that an unregulated market is always superior to regulation. The alternative view, which has been put by some planning commentators and lawyers (Gatenby and Williams 1996), is that the plan-led system is not such a big change in practice and that discretion will remain important in British planning practice. Taking this view implies that the new system represents probably a net increase in planning bureaucracy and no real implementation of a New Right programme.

The proof of this particular pudding will lie in practical experience, and some of the case-study material referred to below is relevant. Bramley and Watkins (1996) provide some statistical evidence that a plan-led system should increase supply responsiveness of private housing.

Turning to changes which, while they may have been mooted have effectively not been pursued, we find some important examples which suggest that planning has been quite resistant to pressures from the New Right. The concept of a general presumption in favour of development was given some support in government guidance in the early 1980s (Circulars 22/80, 22/84). However, this has been effectively overturned by the 1991 Act and subsequent policy guidance. Another obvious line of attack would have been to reduce the scope of various forms of restrictive designation on land, such as Green Belts. Again, there was an attempt to do this in the early 1980s, which was comprehensively defeated by concerted local political opposition (Elson 1986). The government could have acted to outlaw the attempts by local authorities to lay down detailed policies and guidance on 'good design' in relation to housing, policies which gained popularity from the late 1970s. Again, while this may have been something of the flavour of the early 1980s, by the early 1990s the policy emphasis was even more strongly on good design.

Last, but by no means least, the government could have acted to curb the extensive uses which local authorities were making of their powers (formerly section 52, now section 106) to strike planning agreements with developers. Such agreements, and some use of conditions, effectively trade planning permission for certain benefits to the local community generally known as 'planning gain', and these have become increasingly important on major housing developments (Grimley J.R. Eve 1992; Farthing *et al.* 1993). Despite much controversy about the legitimate scope for such instruments (see, for example, Healey *et al.* (1993) and, in the housing context, Joseph Rowntree Foundation (1994)), the government has made little move to curb this practice. Practical examples are discussed later. One can speculate about why planning agreements are condoned, and suggest reasons that show at least some consistency with aspects of the New Right agenda. One reason is that planning agreements often make developments possible and acceptable which would otherwise not be, and developers feel that the cost is worth paying, indicating a probable overall economic gain. A second reason is that they aid the process of cutting public expenditure, particularly local capital expenditure or borrowing by public utilities.

Overall this review of a range of possible approaches to reducing the regulatory scope of planning suggests that on the whole the New Right's ideas have not been implemented here consistently, particularly in the areas that matter most for housing development. Understanding why this is so is a central question for the political economy of planning in Britain.

Reducing the role of local government was identified as another element in the New Right programme. Some highly publicised policy initiatives in the planning field such as UDCs and EZs can be directly characterised in this way, but arguably these were not typical of mainstream planning over the period. Most new housing development was in areas still governed in planning terms by elected local government. Some tendencies have served to weaken local authorities somewhat in the planning task, but overall these may be seen as marginal rather than fundamental. In the 1980s there was an upsurge of planning appeals (see table 4.2), partly due to the development boom, and ministers showed some tendency to let a number of these through in such a way as to undermine local planning policies. This has been less true in the 1990s, due to a combination of lower demand pressure, the plan-led system, and more environmental concern by ministers. A degree of centralisation was also apparent in the readiness of the Secretary of State to increase structure-plan housing requirements. The other significant tendency, particularly apparent in the 1990s, has been the establishment of a system of Regional Planning Guidance, prepared by the Secretary of State, which sets the parameters for strategic and local plans, and in relation to housing specifies dwelling requirements at county level (in England). In addition there has been the proliferation of detailed national planning policy guidance through the PPG series (Quinn 1996). Such guidance must be treated as a material consideration in planning decisions, but is generally open to interpretation in the light of local circumstances.

Planning has been increasingly influenced since the late 1980s by a new *environmental agenda*, but as suggested above the New Right in general may be expected to favour *economic mechanisms* for resolving environmental problems. It is argued that full market pricing is a better way of deterring demand for activities that have adverse environmental externalities than

Table 4.2 Planning appeals received, decided and allowed, total and major housing schemes in England

	All appeals			*Major housing applications*			*Share*
Year	*Received number*	*Decided number*	*Allowed %*	*Received number*	*Decided number*	*Allowed %*	*maj hsg %*
1977	10,833	8,366	29.2				
1978	11,609	8,952	28.0				
1979	12,990	8,933	29.1				
1980	16,208	13,130	30.8				
1981	16,637	14,451	32.7				
1982	13,900	12,915	31.0				
1983	13,699	11,221	32.4				
1984	16,192	11,653	32.4				
1985	17,839	14,130	39.5				
1986	19,856	15,613	39.8				
1987	22,482	18,474	37.6				
1988	28,659	21,061	36.7				
1989	32,381	26,481	33.4	1,781	1,306	31.8	4.9
1990	26,692	26,393	33.6	1,372	1,061	32.8	4.0
1991	22,121	22,553	33.7	1,023	938	33.8	4.2
1992	17,959	17,832	33.5	624	580	35.1	3.3
1993	14,979	14,113	35.3	553	389	31.1	2.8
1994	14,651	12,236	33.8	661	391	37.1	3.2
1995	14,653	11,214	33.5	700	417	35.5	3.7

Sources: Department of the Environment, *Development Control Statistics: England 1995/96*

bureaucratic regulation. Tradable permits (to pollute or develop) are used in the US, for example, so that the market can be allowed to continue damaging the environment as long as market actors are willing to pay. This approach of using market instruments is associated with writers such as Pearce (Pearce *et al.* 1989), who have had substantial influence on government policy statements on sustainable development (e.g. the White Paper *This Common Inheritance*, DoE 1990). This sort of thinking would seem to have more relevance in policy areas other than housing – road pricing, the new Landfill Tax and the proposed carbon tax would be examples. However, there is now some speculation about 'greenfield development land taxes' as a way of deterring urban extensions and encouraging the recycling of existing urban land (see statement by Yeo in *Planning* December 1996).

On town planning more broadly the current policy emphasis, initially set out in *This Common Inheritance* and carried forward in PPGs, is to locate new development in ways that reduce the need to travel, thus reducing the use of non-renewable resources and transport emissions; to use environmental assessment techniques in plans and development control decision-making; and to use planning agreements as a means of mitigating or compensating for environmental loss or damage. All of these measures would seem to place considerable emphasis on technical and professional judgement, and not be consistent with a New Right market-oriented way of thinking. The alternative (deeper green) view would not go along with the view that market instruments and mitigation are enough. This would emphasise ideas such as environmental limits and carrying capacity, i.e. that there are absolute limits beyond which development should not go. Bramley and Watkins (1996) show that environmental capacity plays a significant role in determining structure-plan housing provision levels in some areas, and the issue of environmental capacity is attracting more general interest. The challenge remains, of course, of measuring capacity. This places even more technical demands on planning, and doubts remain about whether all of the relevant factors are quantifiable in planning terms.

In relation to housing there is also some new interest in the idea of 'demand management', arising from the debate over how to accommodate the projected growth of households into the next century (DoE 1995a). One theme is that household formation may be affected by the availability of housing (the 'Circular Projections' argument discussed in Bramley and Watkins (1995) and Bramley 1996), and restrictions on housing supply might induce fewer households to form. Beyond this, the recent Green Paper on *Household Growth* (DoE 1996) discusses a variety of other ways of influencing household numbers. These include finding ways of making it more socially and financially attractive for older people to live with their families or changes to social support or Housing Benefit to encourage younger people to remain with their parents, and reducing the number of vacant houses and more efficient use of existing space. But there are good grounds for scepticism about the scope for such measures. Most emphasis is on issues of location of new housebuilding, with targets for the re-use of urban land for new housing, together with higher densities and intensification. An implication of this debate is that more, rather than less, market intervention may be required – more emphasis on diverting market pressures away from areas of high demand, more use of land acquisition powers or grant regimes to overcome development difficulties in urban areas, more attention to housing type, density and design. This implies a renewed emphasis on the traditional concern of planning to manipulate the geographical distribution of development in order to achieve objectives that override meeting housing demands wherever they arise, and would seem to run counter to the early 1980s policy of facilitating market demand.

Privatisation of state functions is another well-known plank of New Right thinking, but planning has not (so far) been included in the services prescribed for Compulsory Competitive Tendering (CCT), which would have been one obvious route to implementation. The attempt to involve housebuilders in joint land-availability studies enjoyed some success in the early 1980s but was essentially an informal consultation mechanism rather than any transfer of responsibility. Planning consultants are extensively employed on specialised planning studies, research and project development, but there is nothing new in this. New settlement proposals promoted by the private sector are a good example where most of the planning may be in the hands of the private sector, but relatively few such settlements have been implemented.

A more significant feature of the 1990s has been the promotion of partnership in order to lever resources and better co-ordinate development and regeneration activities. Promotional activities have grown in the urban policy field with the introduction of the Single Regeneration Budget (SRB) and the establishment of English Partnerships, and Priority Partnership Areas in Scotland. In the housing field there has been a growth of local housing partnerships involving local authorities and housing associations, a greater reliance on private finance to provide new social housing, and specific partnership projects with housing associations and housebuilders undertaking joint schemes for mixed-tenure developments.

There have also been significant changes in the wider context in which planning for housing operates. Following the housing market collapse of the early 1990s, and a rather uncertain future for continued expansion of owner occupation (Holmans 1995), housebuilders are having to respond to a more demand-driven environment where considerations of cost, quality and diversity may be more important than volume production. This might be expected to take the heat out of the land supply controversies that were such a prominent feature of the 1980s. Nevertheless, land supply constraints and growing opposition to new housing development in some parts of the country continue to preoccupy the housebuilding industry, together with fears that the new 'plan-led' system will fail to deliver sufficient development opportunities. Delays in getting the new local plans adopted in many areas are adding to, rather than reducing, uncertainty.

Planning for housing at the local level

The earlier part of this chapter discussed a number of the implications of New Right thinking for planning for housing. Some of the key implications are: that planning should shift from a directly controlling to a responsive mode of operation; that planning should facilitate private-sector housing provision, responding to private-sector demands; that regulatory costs should be reduced; and that control of the planning process might be handed over to the private sector. To what extent have these implications been followed through in local practice?

Reduction in public-sector control

Public-sector withdrawal from a direct development role, and more reliance on regulatory frameworks and negotiation with the development industry, were a strong feature of change in new housebuilding in the 1980s. The origins of this change, however, go back to the major public-expenditure reductions of the mid-1970s and the shift in emphasis from managed decentralisation of population and economic activity to a concern with urban decline following the 1977 Inner Urban Areas Act. The impact of this change is probably fairly widespread, but its effect is particularly felt in areas, such as New and Expanded Towns, where planned development in the post-war period was substantially led by the public sector, as well as in the major cities where significant municipal housing programmes were undertaken. Recent research on Swindon (Boddy *et al.* 1997) illustrates the nature of this shift and the consequences for planning practice at local level.

In Swindon local commitment to growth and expansion is long-standing. The local council responded enthusiastically to the possibilities for expansion offered under the 1952 Town Development Act, acquiring land through compulsory purchase and negotiation and seeking planning permission for new housing, frequently against the opposition of the county council and surrounding rural interests. Extensive areas of council and mixed public- and private-sector housing were developed in the town during the 1950s and 1960s, and the disposal of land provided finance for social and community infrastructure and a stream of funds for further land purchase. In later phases of expansion, the local authority maintained this central controlling role, drawing up plans for a series of urban villages, and buying up land ahead of subsequent development by the private sector. In the period up to the late 1970s the local authority therefore played a leading role as landowner and developer alongside the private sector.

By the beginning of the 1980s growth pressures in Swindon intensified with the take-off of the M4 corridor phenomenon. Economic relocation to the town continued, but growth was increasingly driven by speculative housing pressures as housebuilders sought to take advantage of strong demand in the locality. This arose from the economic success of the town, but also from housing demand diverted from the overheating South East housing market. By now, however, the context and the role for the local authority were very different. The council's landholdings were nearing exhaustion and it lacked the financial resources to assemble sites ahead of development. It could no longer exercise the degree of control over the development process which, as landowner, it did in the past. The recognition that the authority would have to rely much more on conventional planning powers was one of the factors leading to a reconsideration of the growth strategy and increasing concerns over meeting the costs of growth. Consolidation, rather than growth, was being promoted locally.

At the same time, however, national policy was emphasising a facilitative stance in relation to private-sector housebuilding. Several major housing sites were subject to appeals which were allowed by the Secretary of State, and in 1986 a major application for a further 1,500 acre town expansion scheme (the northern sector) was submitted by a consortium of volume housebuilders, who had been buying up land in the area over the previous years. The scheme was twice the size of some of the well-publicised new settlement proposals of the 1980s, and also represented an ambitious attempt by the private sector to take responsibility for the comprehensive planning and development of major urban expansion. Both the county and the district council were opposed to the scheme going ahead in the form proposed. But, as a departure from the strategic planning framework, the application was called in by the Secretary of State, and following a public inquiry given permission in 1988.

This decision was followed by a long period of complex manoeuvrings and negotiations, which tested the ability of the authority to maintain control of the development process. Very substantial requirements for infrastructure and social provision for the development of around 10,000 houses were to be the subject of a section 106 agreement. Negotiations over the section 106 agreement were protracted for a number of reasons. The initial application, accompanied by a master plan, proposed an exceptional package of supporting infrastructure, but the funding for this was dependent on housing-land prices prevailing at the time of the permission and a high projected rate of housing completions. As the housing market went into decline in the late 1980s and land prices fell, the financial basis of the original deal was severely compromised. One member of the consortium withdrew from housebuilding and pulled out of the scheme, and separate applications for parts of the site were submitted by the remaining consortium members. From the local authority's perspective the danger was that the land holdings would fragment and the prospects for a comprehensive scheme with a full range of infrastructure would disappear. Eventually a solution to this problem was negotiated, involving a phased programme of development, a retention of the original package of infrastructure, and a mechanism whereby land at the margins of development is conveyed to the local authority as each phase of development takes place, so that later phases of development can only proceed when the terms of the section 106 agreement are fulfilled. Maintaining control and ensuring the terms of section 106 agreements are kept to has required considerable ingenuity and a process of negotiation taking almost four years.

Swindon, then, has been forced to make the transition from exercising control via land ownership, to a reliance on narrower planning powers and negotiated agreements. Its ability to do this reflected the strong market for development in the locality in the late 1980s, though as the market declined preserving the terms of the original agreements required considerable resourcefulness on the part of local authority officers. In achieving this outcome skills acquired as a result of the local authority's own past development activities were significant local resources.

Facilitating the private sector

That planning should facilitate private-sector housebuilding through its land release policies was probably the clearest message being conveyed to local planning authorities during the 1980s. Where local policy-makers were reluctant, then central government powers to modify structure plans or give permission via appeals could be used.

Historically, planning for housing has been based on the principle of urban containment (Hall *et al.* 1973), an objective that the planning system has pursued with some success. However, private-sector housing has until very recently been concentrated on greenfield sites on the edges of towns and cities, often leapfrogging the Green Belts that were established around the major conurbations. Consequently, it is perhaps more accurate to interpret the outcome of planning policy as one of managed decentralisation of population and employment. More recent research suggests that the planning system in the early part of the 1980s continued to pursue managed decentralisation, together with the protection of valued areas of landscape and open space around large cities, with some success (Healey *et al.* 1988). Other studies tend to confirm that policies of restraint were upheld during the 1980s, certainly in parts of the South East where local opposition to growth remained a strong feature. Cheshire and Sheppard (1989) and Evans (1987, 1991) argue with some empirical support that planning constraints in southern high-demand areas resulted in levels of housing output below potential market demand, accompanied by somewhat higher house prices and higher housing densities. Evans interprets this as in part a response to local political opposition in many parts of South East England, and the power of the rural lobby and suburban residents in planning at a local level. The spread of owner occupation may well have contributed to this, as more voters acquired an equity stake in their local environment and most new development was likely to be perceived as having a negative impact on house values.

The land supply issue, such a prominent feature of debates in planning for housing in the 1980s, also has a longer history, first emerging in the early 1970s, as owner occupation grew in significance and private-sector provision came to dominate new housebuilding. Growing controversy and lobbying by the housebuilders led to a number of studies of land availability and planning (EIU 1975; JURUE 1977; DoE 1980a; Tym 1990) which shed light on the process of housing development and assess the role of planning in constraining output. These studies reveal a good deal of continuity in the process and in the impact of planning. Planning constraints are important in some places (particularly areas of high demand in the South) and at some times (when market conditions are buoyant), but they also demonstrate the significance of other factors in constraining output, notably uncertainties surrounding market demand, provision of infrastructure and the behaviour of landowners. The findings also demonstrate the flexibility of the planning system in that period: a relatively large amount of development took place on land that was not formally identified in plans, and the release of such 'windfall sites' was an important way of maintaining responsiveness to market demand. In Britain plans have always been indicative and only one of the considerations taken into account in making development decisions. Flexibility is therefore a feature of the planning system that pre-dates the era of New Right thinking.

A more systematic attempt to model the effect of planning on housing output, prices and density during the mid-to-late 1980s and early 1990s is represented in Bramley *et al.* (1995) and Bramley and Watkins (1996). Some of the key findings of this work are that planning does have the effect of constraining housing supply below what the market might be expected to deliver, but that the output and price effects are rather modest. While planning, at the level of setting targets or giving planning permissions, is not especially responsive to market demand, it cannot be characterised as totally rigid; flexibility is maintained through the release of 'windfall sites' as discussed above. Another finding of this research is that planning tends to concentrate development in certain areas. Growth areas in the 1980s include a number of particular localities on the fringes of the South East; other areas within the metropolitan Green Belts and the fringes of historic towns and cities had lower growth, reflecting very tight planning constraints. Many of these constrained areas are characterised by high market demand and high house prices. This suggests that restraint policies have been largely upheld in the most environmentally and politically sensitive areas, and that planning has acted in practice to steer demand to particular localities, where perhaps local opposition is (at the outset) less. Such a strategy may also have unintended and potentially undesirable impacts on the type and mix of housing provided. A study by Monk and Whitehead (1996), for example, found that releasing large amounts of land in lower-demand areas of Cambridgeshire resulted in supply being concentrated on very small units for the first-time-buyer market, not something the planners entirely favoured.

One interpretation of planning for housing in the 1980s is therefore that policy relied substantially on a small number of 'safety valves' away from the areas of highest demand and highest constraint. Towns such as Northampton, Peterborough, Milton Keynes, Wokingham and Eastleigh saw high levels of housing completions. In the growing M4 corridor new housing development was highly concentrated in a small number of localities. In the counties of Avon and Wiltshire, for example, high projected housing requirements were accepted by the local authorities (in both cases following Secretary of State intervention to increase structure plan targets to capitalise on the growth potential of the areas), but land allocations were highly concentrated on a relatively small number of large housing sites. In Avon, Bradley Stoke to the north of Bristol (planned to accommodate 10,000 new houses) was a key strategic land allocation located in an advantageous position relative to the motorway network, but also under a local authority politically pre-disposed to the more relaxed planning regime being advocated by central government. In Wiltshire, new housing allocations were concentrated in Swindon, again reflecting its favourable location on the motorway, in an authority with a long history of supporting growth and expansion. The large northern sector scheme was, however, the subject of an appeal decision, overriding the opposition of local interests to further growth on this scale. In both cases new housing development went along with significant economic development, reflected in new manufacturing and business park developments and substantial growth of service-sector employment. Both Swindon and North Bristol represent examples of a more dispersed form of urban development that was characteristic of the 1980s ('edge cities' in Garreau's terms), located in corridors of growth extending out of London (Hall 1995).

In a strategic sense, therefore, planning in the 1980s facilitated the market, but in a rather selective way. In some cases this was implemented following central government intervention to increase housing requirements through structure plans or appeal decisions; in other places it reflects a continuation of long-standing policies of population and employment decentralisation to strategic growth points in and beyond the South East. There is no evidence, however, of a widespread process of deregulation. Planning constraints in many areas have been upheld, partly through policies of selective concentration and diversion of housing demand to less contentious localities. Significantly, none of the controversial new settlement proposals promoted by housing developers in the South East in the 1980s received support from central government. The local politics of growth and development, a feature of planning from its inception, remained a significant factor during the period.

One finding of the Bramley and Watkins (1996) study was a noticeable tendency for areas which previously experienced a high level of new housing development to be reducing their planned provision for new housing. This is illustrated by the contemporary debates about 'consolidation' (in Swindon) and 'capacity' (in Bristol and other areas in the South). This reflects growing local opposition to a continuation of past policies, which have concentrated growth pressures to a significant extent, and the higher salience of debates on sustainability and environmental capacity. It also raises issues about the ability of the planning system, in the future, to deliver an adequate overall supply of housing in a context where projected levels of housing need continue to be relatively high. In a 1990s context the problems of local political opposition to development are becoming more widespread, and current discussions suggest an intention that existing urban areas become the new 'safety valves' (Breheny and Hall 1996). The focus for resolving growth pressures is now shifting away from structure plans to the level of Regional Planning Guidance which has become the key means for central government to require localities to meet projected housing needs.

Reducing the extent and cost of regulation

One justification for the introduction of the plan-led system is a reduction in uncertainty and fewer costly appeals and delay at the development control stage. A difficulty in assessing whether the plan-led system will deliver these benefits is that in many areas the new local plans have been subject to considerable delay at the preparation stage. This reflects an increase in lobbying and involvement on the part of landowners and developers due to the higher status of plans, as well as problems in administering the volume of plan-making activity under way. In areas affected by local government reorganisation, political and administrative difficulties continue to affect the new arrangements for structure planning, and may be adding to uncertainty. So, although in principle the plan-led system may be consistent with providing certainty and less delay, in practice the reform is having the opposite effect.

In Wiltshire, for example, a long-running battle between the county and Thamesdown District over strategic land-allocation policy is being played out in the context of a replacement structure plan for the area. The county policy of concentrating growth pressures and land allocations in Swindon is resisted by the district, which in April 1997 achieved unitary status and became a separate structure-plan authority. It is very unlikely that the district will agree to adopt the county-prepared structure plan. Meanwhile the local plan for the town runs out in 2001, and current housing land allocations in Thamesdown are insufficient to meet the proposed structure-plan requirement. In the absence of an agreed strategic framework it is not possible to roll forward the local plan. A protracted period of uncertainty looks set to ensue, that may only be resolved by central government intervention to impose a structure plan. In the meantime housebuilders are lining up options on land in the area, and threatening appeals if the planning system fails to provide development opportunities.

In two other important respects we could not conclude that changes to the planning system in the 1980s have reduced the costs of regulation.

One concerns the use of section 106 (formerly section 52) powers to require contributions from developers for infrastructure and supporting services in new housing developments. The other concerns the use of planning powers to require a proportion of 'affordable housing' as part of market housing schemes.

Both the Bradley Stoke development, north of Bristol, and the northern sector expansion of Swindon, involved the developers in very substantial packages of supporting infrastructure. Other research confirms that this practice grew during the 1980s (Barlow and Chambers 1992; Barlow *et al.* 1994; Farthing *et al.* 1993; Healey *et al.* 1993). While negotiations on both these sites were protracted, and also contentious in relation to some detailed aspects, the development consortia involved in both developments accepted from the start that infrastructure costs would be borne by the development. The acceptance of this was one reason for selecting these very large developments, to be undertaken by a number of development companies working jointly, though in both cases problems of enforcing the terms of initial agreements have followed from the subsequent recession in the housing market. There seems no reason to expect that this policy of loading infrastructure costs on to developers will be reversed in the future. There is, however, an emerging contradiction between policies for sustainability that emphasise recycling urban land and pressure on the house-building industry to meet the costs of social and community facilities. The smaller scale and higher cost of development within urban areas may not generate the kinds of surpluses required to fund substantial packages of planning gain, though presumably some infrastructure costs would be lower in such locations.

The second feature of change is the new emphasis on the role of planning in enabling the provision of social housing. The policy originates from local experiments with requiring social housing as part of negotiated planning agreements during the 1980s in the context of growing affordability problems and severe constraints on local authority spending on housing. In an apparent turn-around in central government views about the appropriate range of planning powers with regard to new housing development, the government accepted in 1989 that the need for affordable housing was a material planning consideration in rural locations, and then subsequently incorporated this principle in more general form in PPG 3 (DoE 1992). Authorities are now encouraged to negotiate with developers seeking planning permission for market housing for inclusion of an element of affordable/social housing, with free or subsidised land or completed units being made available to housing associations. This change further extends the range of planning gain that authorities are allowed to seek, and encourages negotiation on a site-by-site basis. It also poses new policy implementation challenges, in that detailed attention to the specifics of the local housing market and to the economics of development on particular sites is a pre-condition of successful negotiation of affordable housing quotas. Despite these difficulties many local authorities, particularly in the South of England, have taken up the challenge of trying to exploit these new possibilities (Barlow *et al.* 1994).

Private-sector planning

As we suggested above, there is little evidence that significant parts of the planning function have (yet) been handed over to the private sector. Brindley *et al.*'s (1996) example of 'private management planning' for the regeneration of a run-down local authority estate was a short-lived experiment that in the end required substantial public-sector subsidy to achieve its objectives. Similar experiments elsewhere (for example, the Thamesmead Estate in London) may have had more success, and the general idea has been carried forward in the Housing Action Trust initiative, though take-up has been relatively low and the need to commit substantial public-sector resources makes these initiatives more like partnerships than pure private-sector initiatives. The new settlement proposals promoted in the 1980s are an example of where most of the planning was carried out by the private sector, but few such settlements were implemented in the face of substantial local political opposition. Current government guidance on new settlements makes clear the need to bring forward proposals through the local plan process.

However, a number of the large-scale new residential developments taking place in the 1980s did employ a model of planning and implementation that changed the respective roles of the public and private sectors, as discussed in a previous section, implying a more significant co-ordinating role for the development industry. Bradley Stoke on the edge of Bristol and the Swindon northern expansion are both examples of major housing developments where implementation was carried out by development consortia, taking responsibility for detailed master planning, and funding infrastructure provision from profits derived from selling land on to housebuilding companies. This is essentially the new settlement model that was being promoted elsewhere, but the problems of such a model are illustrated by these developments. The ability of the private sector to co-ordinate development and effectively provide new infrastructure in line with housing development may be undermined by the instability of land and property markets. Bradley Stoke, where development is well under way, but at a much reduced rate during the recession, has been the subject of extensive criticism and a sustained campaign by new residents over the lack of social and community infrastructure. In Swindon the recession threatened to undermine the original commitments made by the developers in a more favourable market context, and rescuing the scheme required complex and protracted negotiations.

Continuity or change in planning for housing since 1979?

There is much analysis of public policy during the 1980s that suggests more continuity than change. The reasons may be due to inertia, political pressures and unexpected events, as well as contradictions within New Right thinking, which as we discuss may be particularly significant in the case of housing. The discussion above suggests that the implementation of the New Right agenda in planning for housing was less than consistent. And others commenting on broad changes in the nature of planning also find evidence of more continuity than change (Griffiths 1986; Reade 1987; Healey 1993).

The general thrust of these arguments is that, while the government in the 1980s frequently expressed hostile sentiments about planning, it was neither feasible nor desirable to dismantle planning controls, giving rise to the mixed picture described in the previous sections. First, market processes have always dominated the land and development process, with planning having a complementary regulatory role – aiming for efficiency in the use and provision of infrastructure, protection of landscape and the countryside and balanced housing and employment development (Bramley *et al.* 1995: ch. 3). Without planning, land and property markets are prone to failures, under-supply or inefficiency in the provision of infrastructure, under- or over-supply of particular kinds of property, or the generation of adverse externalities. According to this analysis, market actors are not averse to planning controls that preserve the value of land and property and provide greater certainty. Second, it is argued that beneficiaries of planning controls have always been more middle-class suburban and rural interests, which a Conservative government would want to continue to support. The maintenance of restraint policies in many high-demand areas of the South during the 1980s, alongside policies which concentrated land release in a selective number of areas, suggests that the interests of rural and suburban residents remained a potent force during the 1980s. Current policy guidance is, if anything, reinforcing this view.

Nevertheless, some changes did occur. Public expenditure restrictions undermined the scope for positive planning through infrastructure investment and public-sector land acquisition and assembly. Thus incentives for private development to follow 'plans' were less, and mechanisms for steering the market less clear. One consequence was the onus on planners to negotiate with developers to secure infrastructure in the form of 'planning gain'. In the course of such negotiations it is possible that certain standards may be traded, that accountability is reduced (the need for confidentiality), and that certain kinds of development are favoured (e.g. large-scale development from which more 'gain' may be extracted). Also, as new spatial divisions emerged, consequent on economic collapse in some areas and growth in others, the effectiveness of universally applied negative controls was undermined. Hence the need to explore different forms of planning in different areas (Brindley *et al.* 1996). Finally, it is claimed that a greater emphasis on a negotiative form of planning, together with increasing diversity in land and property markets, required planners to become more 'market-sensitive' (Healey 1993). The flexibility of the planning system allowed for a greater accommodation of development interests, without any significant reform. While there may have been a shift in the relative power of different interests, there was no fundamental change in the shape of the planning system (a hierarchy of plans and comprehensive development control). Indeed it might be argued that a better understanding of market criteria and the operation of land and property markets is a strength rather than a weakness.

The re-assertion of the importance of planning resulting from the 1991 Planning and Compensation Act's promotion of a 'plan-led system' has been identified as a key shift affecting housing. This was a response to a combination of factors – the growing strength of environmental concerns, the political fallout from the development controversies of the 1980s (especially in the South East), and an acknowledgement of the damaging effect of 'overbuilding' at the end of the late-1980s property boom. On balance we interpret this shift as a reinforcement of planning regulation rather than a net reduction.

More fundamentally, the environmental sustainability agenda, rather than emphasising the New Right's favoured economic mechanisms, is reinforcing the case for regulation in relation to a number of aspects of new housing development. It implies amongst other things a shift in land release policies to favour urban sites and higher densities; to the extent that long-standing policies of urban containment are reinforced this policy is not a complete departure.

From our initial review of New Right thinking and its implications for planning for housing, we drew out six major elements of what would have constituted a thoroughgoing New Right policy programme. Of these six, only one has been unambiguously and consistently implemented in this period – the reduction in the direct role of the state in housing provision and the consequent loss of the option of direct state-led development. Facilitation of private housing development featured strongly in the rhetoric and in policy initiatives of the early 1980s, but taking account of more recent changes and evidence on how the system actually operates, the picture remains very mixed, with facilitation happening at best on a selective basis. Planning regulation has not been substantially dismantled; some streamlining and by-passing have been achieved at the margins, but in the core area of control of new housing development regulation has on balance been entrenched with the 1990s move to a 'plan-led system'. The role of local government has been curbed or overridden to some extent, but local planning authorities still retain a crucial role in steering, controlling and negotiating new housing developments. Environmental issues are increasingly influential in planning policy, and it is noteworthy that these are expressed more through a regulatory mode than through a fiscal or market-oriented mode, again contradicting the prescriptions of the New Right. Finally, planning for housing has not been substantially privatised, although the consortia of developers involved in major new schemes do represent a partial example of this phenomenon.

Why has the New Right's impact on planning for housing been so modest, relative to its overall agenda? Why has regulation been entrenched rather than uprooted in this arena in the 1990s? Three main reasons seem to be particularly worth highlighting. First, the fundamental economics of land development remain as always subject to significant potential market failures associated with externalities, local public goods, infrastructure and uncertainty; planning is used and often valued by most participants in the process as a way of coping with these problems. Second, the political strands of the New Right include two tendencies, the liberal and the traditional, which entail significant contradictions in relation to the issues raised by planning for housing, with the traditional conservative stance legitimising the anti-development stance of rural landed and suburban homeowning interests. Third, a pragmatic and populist political style has reinforced this tendency, recognising the unpopularity of development in many sensitive areas for example, or opting to respond to environmental issues through a regulatory rather than a fiscal approach.

5

DEVELOPMENT PLANS: COPING WITH RE-REGULATION IN THE 1990S

Angela Hull and Geoff Vigar

Introduction

This chapter assesses the influence of localities over development plan-making in the 1990s in the aftermath of Conservative governments' rhetorical attack and system fine-tuning in the 1980s and early 1990s. The impact of this system 'reorientation' is appraised and aspects of consolidation and change under the governance of John Major are identified. Our conclusions regarding the influence of localities are primarily drawn from detailed case studies of Lancashire, West Midlands and Kent, carried out as part of research sponsored by the Economic and Social Research Council (ESRC).

We first draw out the contextual changes for planning, in terms of the legal powers and responsibilities given to different agencies, government advice to local planning authorities (LPAs) about appropriate issues to consider when devising plans, and the role of central government in monitoring local authority (LA) decisions. Broader ideological and rhetorical pronouncements by central government also influence the rationale for planning intervention in development decisions and may influence the stance of developers and third parties too. We compare the approach of Thatcher and Major administrations to the town planning service in an attempt to understand both the purpose of central government efforts to structure decision-making, and the clarity of the message received. Later in the chapter we assess local responses to central direction, concentrating on three aspects of plan-making: the driving forces of development plan production, procedural aspects, and the centralisation of policy formulation.

The development plan framework: 1979–90

Through the 1980s functions were lost by local government, budgets were cut back and even some geographical areas were removed from the aegis of

LAs. Major changes occurred in development plan-making through a combination of strategies to change both the climate and the purpose of planning (Grant 1990; Rowan-Robinson and Lloyd 1986; Thornley 1991). We group these changes into three main categories.

First, the government exerted its control through the legal system where it was less likely to be challenged. The 1980 Local Government, Planning and Land Act sought to reduce the strategic powers of county councils and to limit the scope of the planning exercise. This was achieved by allowing local plans to be prepared and adopted prior to the approval of a structure plan, thus breaking the rigid hierarchical nesting of plans whereby the level above provided a framework for the level below. The Act also reduced the implementation role of the counties through giving greater development control powers to District Councils, even though these powers had to be exercised so as to achieve the general objectives of the structure plan. The requirement to consult on the strategic plan issues was removed, as was the need for a public local inquiry where objections could be dealt with through written representations. The Secretary of State for the Environment (SoS) no longer required all development plans to be approved by him in their totality. The 1986 Housing and Planning Act brought forward expedited procedures for the preparation of local plans in non-metropolitan areas, essentially reducing public participation and consultation.

Second, the government's representation of what it saw as the role of the development plan was conveyed mainly through government circulars, statutory instruments, and Department of the Environment (DoE) scrutiny of plans at inquiry. The DoE thinking at this time was that plans should be low-cost, speedily produced schedules of land availability in areas under pressure for development. Specifically they should identify locations for the future supply of land for housing and industry, define the precise boundaries of areas of restraint, and/or co-ordinate programmes for development (DoE 1981). Issues of resource availability and equity were superfluous for plan-making, because plans were seen as tools to *assist developers and the business community by providing them with some indicators to guide them in taking their decisions* (DoE 1985: 14). This new representation of the plan was conjoined with unprecedented (for planning) powers of monitoring, scrutiny and intervention for the Secretary of State during the process of plan preparation. The 1986 Act laid down powers for the SoS to direct a local planning authority to make, alter, repeal or replace a local plan. These two acts centralised legal control to such an extent that:

> One of the most distinctive trends of the past nine years has been the emergence of Government policy as a dominant force in development control. It has been brought about by a series of hard hitting circulars based on the Government's deregulatory and

> pro-development ideology, coupled to a new willingness to use the appeals process as a means of reinforcing its policies.
>
> (Grant 1990: 143)

Third, the government used its political power from the outset to implement its conception of planning and the role of the public sector. It frustrated local authority political ambitions through cuts in public spending, the abolition of the metropolitan counties in 1986, and through chipping away at the more restrictive procedural aspects of plan-making. The government's view of the problem was of planning regulation being cumbersome, costly, ineffective, and misunderstanding the development market (DoE 1985). More extensive reform of the development plan system was subsequently proposed and crystallised in a White Paper, *The Future of Development Plans* (DoE 1989a). This document proposed district-wide development plans for the non-metropolitan areas with a statutory duty to review them regularly and keep them up to date. More controversial was the suggestion to 'abolish' structure plans and to replace them with statements of county planning policy. The White Paper also announced a 'new' hierarchy of planning documents to include regional planning guidance notes (RPGs), planning policy guidance notes (PPGs) and mineral policy guidance notes (MPGs).

Thus, we see that by the end of the 1980s the rationale for comprehensive, locally derived planning was under attack. A more interventionist role was prescribed for the SoS through the increase in appeals, the use of circulars, the preparation of regional guidance notes, and the introduction of additional statutory reserve powers. At local level the strategic policy-making role of the County Councils had been emasculated through abolition or dilution of powers to District Councils, and opportunities for local authorities to legitimate their emerging policies had been reduced. The 'discretion' to structure local agendas, through imposing material considerations and a more prescriptive role for guidance, had reverted back to the SoS.

The formal planning framework was still intact at the end of the Thatcher administration. However, the imposition of new values and goals had brought into the open the very contradictory nature of town planning intervention (Healey 1983) and had questioned both the effectiveness of that intervention and the expertise of the planning profession. Planning to provide for community needs and to secure an environmentally efficient arrangement of land uses, was being downgraded at the expense of providing a more supportive environment for private-sector developers and some degree of certainty for market players.

The success of this deregulatory atmosphere in the mid-1980s had brought about over 200 proposals for new settlements in the countryside, and many out-of-town retail and business developments. However, especially with regard to new settlements, neither the predominantly Conservative councils affected, nor the government, have consistently backed developers in the face of hostile residents. Support for environmental protection from within the ranks of the Conservative Party has called forth a more spatially differentiated response from the government towards the planning control mechanism.

Re-regulating planning under Major: 1990–96

John Major inherited a planning system in the process of reform and a planning policy community confused by new possibilities for managing future land-use change and unsure of their role and that of the development plan in guiding development. Most of the intent of the 1989 White Paper was inserted into the Planning and Compensation Act 1991 which was an attempt to provide universal coverage and a clear framework for the determination of planning applications (MacGregor and Ross 1995). Structure plans were retained, whilst responsibilities were allocated for the preparation of the district-wide local plan, the minerals local plan, and the waste local plan. These plans did not require the SoS's consent and could be amended without recourse to the previous lengthy procedures. Unitary development plans, structure plans and district-wide local plans were to continue to include policies for: 'the conservation of natural beauty and amenity, the improvement of the physical environment and traffic management. They must have regard to national, regional and strategic planning guidance, and to the resources available' (Schedule 4, Planning and Compensation Act 1991).

Although this apparent commitment to planning might appear to be in contrast to New Right attitudes, it is clear that creating certainty for the market remained a key element within this new prominence for planning. In addition political tensions resulting from pressure for development in the South East of England and the high-profile failure of deregulated planning in areas such as London Docklands had exposed inadequacies and conflicts inherent at the heart of the New Right agenda.

Section 54A

Any confusion about the usefulness of the development plan was laid to rest with the insertion of section 54A (S54A) into the 1990 Town and Country Planning Act, stating that planning decisions should be taken 'in accordance with the plan unless other material considerations indicate otherwise' – in many ways a statement of continuity, but seen as a triumph for planning at the time. It has been suggested that this action by the government was ill-conceived: a rash response with little thought given to the role and meaning of development plans and less thought given to the legal problems that might arise in the drafting and tinkering with the planning system (Herbert-Young 1995). Some indication of government intentions can be imputed from Sir George Young's opinion of the impact of S54A:

> [the] major shape of development in a district will therefore be determined earlier in the process – as the plan is prepared. This underlines the importance of public participation. . . . If the SoS does not intervene, when he has the opportunity, users of the plan can take it that he would be content to give the plan substantial weight on appeal.
>
> (Quoted in Herbert-Young (1995: 299))

S54A therefore potentially increases the status of the development plan as an arena and mechanism for expressing public preferences for spatial change. Central to this is the ability of various 'publics' to make representations at pre-consultation, deposit and inquiry stages, to prompt intervention by the SoS, and to seek court orders quashing modifications. One of the unintended outcomes of the policy action, borne out by recent experience, is that:

> Greater involvement, together with greater reliance on experts, will almost certainly mean greater delay in plan production, which is contrary to the requirement set out by both the Government and the courts that plans be up-to-date, specific and relevant.
>
> (MacGregor and Ross 1995: 55)

The presumption in favour

Evidence of the lack of coherence in government advice to planning authorities is provided by the inconsistency in approach between S54A and PPG 1 (DoE 1992g: para. 5) which continues the long-standing presumption in favour of development where there is no demonstrable harm to interests of acknowledged importance. There would therefore appear to be many caveats to the presumption in favour of the development plan. First, the LPA must be seen to have taken account of regional guidance, current national policies and ministerial advice. Second, the plan must be up to date and have the support of local people and businesses. Third, there must have been no change in central government advice relevant to the proposal since the development plan became extant. The majority of the judicial interpretations of S54A have hung on whether the offending development proposal will have an adverse effect upon the purposes of the plan in general, rather than the literal application of every relevant policy (Herbert-Young 1995).

The rationale for S54A can now be seen as a prompt to developers and their agents that the local plan is the arena in which to influence and inform the local authority of land-market and commercial needs. In part, S54A has hampered market opportunity and led to an increase of power to the LPA, especially given that the plan can be self-adopted. In 1993 the SoS, George Young, felt the need to remind LPAs of the key role they play in stimulating economic recovery: 'it is increasingly local authority planners who are the true enablers. Through your development plans you set the scene to enable economic development to take place' (DoE 1993e).

This element of 'local choice' in the 'decentralised authoritarian' approach by central government is however heavily circumscribed by criteria specified in national guidance and by the 'competitive economic environment which often determines local political priorities' (Thornley 1996: 7).

Development plans: a new agenda?

Major's administration has had the task of ensuring that the new, more flexible plan arrangements could be realised. The first target date set for the comprehensive coverage of land-use plans was 1996. By September 1994 only 10 of the 70 unitary authorities and 65 of the 296 non-metropolitan authorities had taken their plan proposals through the 'streamlined' development plan procedures to adoption (*Planning* 1995). A DoE consultation paper (DoE 1994e) urged LAs to remove much of the detail from plans and to use the six weeks consultation period on the deposit plan more effectively. More strategically it instructed LAs to prepare early for the public inquiry, to negotiate effectively with objectors, and to set a firm agenda for the inquiry. Recently fears have been expressed by practitioners that despite the cost and effort expended, there is a danger of local plans ending up as 'anodyne statements lacking any real detail . . . where important decisions are made outside the process' (*Planning Week* 1996). How far planners have become the 'new enablers', and the response of LPAs to their new role in producing development plans, are discussed in the next section.

Assessing local response to central direction

The scope for local-level implementors to deflect the intentions of central government policies, through their capacity to bend the policy means to achieve their specific local interests, has been well discussed in the policy literature (Barrett and Fudge 1981; Hjern and Porter 1981; Rhodes and Marsh 1992; Ham and Hill 1993). Our research for the ESRC has looked at evidence of local agency in using plan procedures and in formulating the development plan. Questions of access have been explored in terms of who gets involved and who is excluded. We examine the stores of knowledge judged important in defining policies, identify the arenas in which issues are debated, and assess the impact on local stakeholders. The research project covered a 24-month period through 1995–7. It focused on how the key actors in a region inter-relate, what they construe as the main strategic issues, and the place of the planning system in resolving these. From this we gained a feel for the discourses on spatial change in each region as interviewees identified the influences upon them to act. A later research stage attempted to reconstruct the network of relations within each theme and to locate the planning system in a broader regulatory environment. The research provides a wealth of information to throw light on some of the development plan issues raised in this chapter.

In the material that follows we focus specifically on the development-plan function looking at the process of preparation, the advice and intervention from central government, and other driving forces which impact upon the form and scope of the plan. We use these three issues to explore questions of structure and agency in English development plan-making. The specific questions we ask of our data are discussed below.

Plan driving forces

There are many forces that lie behind the strategic direction of development plans. We look at how perceptions of economic and social forces are affecting plan content and direction. However, we choose to focus in particular upon environmental criteria as a driving force in plan production.

Concern for protecting present valued environments and the ecological impact of development on air, land and water resources have been given greater political currency through European Commission directives on environmental quality. The government has sought to implement these partly through the planning system by 'add-ons' to plan scope such as requiring that a section on environmental sustainability is included, and also by establishing mechanisms for environmental impact assessment of development proposals with significant environmental impacts. A parallel system of environmental remediation, based on the polluter-pays principle is being developed and spearheaded by the remodelled Environmental Agency. Government messages on environmental protection policies have been presented via a 'drip-feed' of PPGs (PPG 1, 12, 6, 13, 23), pollution emission standards, small-scale local-level initiatives (Local Agenda 21, Going for Green), some fiscal measures (landfill tax, fuel tax), and voluntary producer initiatives (recycling packaging, etc.). It will therefore be of interest to discern the clarity and influence of the message in each of our three case-study areas.

As the implications of the 'new' environmental criteria, emphasising damage to air, water and ecological quality, work their way through areas of public policy, it is becoming increasingly clear that what is required is radical governmental intervention in spatial investment patterns and transportation modes. The cleavages within the Conservative Party have exposed the Major government's environmental/economy conflict, and 'conservative' support for protection of their immediate environment has halted the progress of market-oriented deregulation. Concerns for the performance of local economies have to be considered alongside 'environmental' factors. The tensions inherent between these two concerns as they work themselves out in places can be difficult to resolve and are ultimately political choices. The ways in which environmental concern is providing a new agenda for plans, and how these political choices are made, is explored in our case study material.

Plan-making processes

The main point to reiterate here is the introduction in 1991 of the requirement for all district-level planning authorities to prepare district-wide land-use plans. This in itself created resource problems for many of them that had small forward planning functions and no experience of undertaking an exercise of such magnitude. Despite the DoE calling for speedily produced plans, the response from local authorities has not been encouraging for the government. Less than half of all district-wide plans and 30 per cent of Unitary Development Plans (UDPs) were not expected to reach adoption by the end of 1996 (Blackhall and Graham 1996). In this context we need to see how the exercise has been framed, e.g. by a wide range of local interests or by concerns with efficiency and speed.

This raises questions regarding the influence of representations received by the LPA on policy outcomes and the implications for local democracy and planning for the public interest. Whilst it is difficult to determine the level of influence a developer may have had prior to the production of the draft plan, interaction in the formal consultation arenas can be identified. The most active participants in these visible arenas are environmental groups and those within the development industry (Barlow 1995; Adams 1994; Webster and Lavers 1994). With the advice in PPG 12 (DoE 1992e) that effective consultation should take place early on in the plan process before the minimum six weeks consultation, the LPA can decide how to structure this interaction. Large landowners as well as 'developers' with projects have an advantage here in that they will be seen by the LPA as key plan implementors.

The centralisation of policy formulation

Organised residents' groups can make it difficult for developer interests to initiate change, with the effect that developers often prefer to short-circuit local planning arenas and 'have a discussion at a central level' (Albrechts 1995). Calls for a corporatist arrangement on policy formulation at the higher tiers of the planning system have been made by producer groups and specific targets for minerals and housing production are evidence of such arrangements. Important questions to ask are how key producer groups engage with policy-making, at what administrative level of the planning system, and with what effect on open and equitable outcomes.

Of key interest to this chapter is the role of the SoS in ensuring that national policy is carried out. The government offices in the regions have a measure of discretion over how they use their statutory delegated powers of intervention, particularly in terms of the timing of objections. They have been criticised by planners for their lack of consistency and unfamiliarity with new policy areas (Hull *et al.* 1994). Evidence suggests that they meticulously reword and delete policies, often after the plan preparation procedures have been completed, either because policies were too detailed, involved non-land-use issues, or because the wording was considered to be too negative (Rosen 1993; Hull *et al.* 1995; Long 1995; Jones 1996).

Kent, Lancashire and the West Midlands: the influence of localities

We use our research data from three areas to assess how the policy changes described above have impacted in different places and how localities have reacted in formulating and implementing land-use policy. The case studies present discourses about urban-region spatial organisation which identify the underlying power relations and the institutional capacity to interlink economic, environmental and social factors and different stakeholder groups.

Kent, Lancashire and the West Midlands provide differing contexts in terms of institutional arrangements, economic history, division of labour, cultural traditions, political alignments, and spatial and physical form. Yet recently, emphasising the discretion still inherent for LPAs, they have all looked to the European mainland when organising ideas for their spatial strategies and inward investment. To overcome its perceived peripherality, Lancashire has chosen a strategy which emphasises connections to Europe, with routes and nodes where the mainland flows to Lancashire and vice versa. Kent's conception of itself as being part of a Euroregion but fearful that people and investment could flow through a corridor across the county, has led to a strategy designed to capture investment from this corridor. In the West Midlands the conurbation and the region sees itself as part of a European network. In spatial terms this builds upon a continuity of direction for spatial strategy and plans stemming at least from the 1982 West Midlands Structure Plan. At regional level, both the West Midlands Forum of Local Authorities and the North West Regional

Association of local authorities have become increasingly important in developing strategic approaches, for example the provision of large sites for inward investors. Kent has, until recently, principally been pursuing such supra-county matters through the Euroregion arrangement with Nord-Pas-de-Calais, and the regions of Belgium.

All our areas have contrasting views of the local. Lancashire is imbued by the urban Labour council service mentality, seeing itself as the political control centre for the area. Holliday's (1991) research on the rise of the 'New Suburban Right' in Kent County Council (KCC) in the 1980s, suggests that KCC considers itself as one of many powerful institutions in a pluralist local environment. In-depth interviews inside the two local authorities reveal these differences in terms of member–officer interaction also, with members exercising a strong hand in the Lancashire case, whilst Kent has a more partnership style of operation. These differing value systems make a real difference to the shape of local governance and in turn the places being governed. Local authorities in the West Midlands show a variety of organisational styles although the organisational power of officer groupings at the conurbation and regional level is significant here.

Kent

The Kent Structure Plan (KSP) third review is awaiting adoption at the time of writing due to continuing disagreements with the DoE and Government Office for the South East over housing land allocations. This plan updates that adopted in 1990, and provides a planning framework to 2011. The main difference between the KSP of the 1980s and that produced in 1993 lies in the dominance of an environmental discourse. Indeed the speed with which the third review has been instigated may reflect the rise to prominence of such issues in the county. Most of the districts in Kent are proceeding slowly toward the adoption of their district-wide plans; Ashford Borough Council is notable in taking its plan straight to deposit.

In Kent a large number of strategic developments are taking place. The two main ones at present are the Channel Tunnel Rail Link (CTRL), and the Thames Gateway. Both of these present Kent with a number of opportunities and threats. We have tried to explore to what extent development plans reflect or frame these strategic developments.

Plan driving forces

The protection of Kent's countryside has been a major concern for planning policy in the county for some time. The rise of biospheric concerns related to air and water resources and environmental carrying capacity is a more recent issue for the KSP. In economic terms the development opportunities arising from the Channel Tunnel and associated infrastructure provides both economic opportunities and a potential threat for the county's valued environment. Kent planners described the county as shifting from being a 'peninsula' to being a 'corridor' and there was a feeling in some quarters that the county could be 'over-run' (planner). Thus an environmental threat alongside the economic opportunities associated with the tunnel and the high speed rail link was highly pertinent: 'how many people do we let in?' (planner).

In strategic terms the county loosely divides into two areas, one where the protection of the countryside is the overriding principle and the other where the creation of employment opportunities is paramount, within certain environmental and planning constraints. This reflects socio-economic conditions in the two areas. Thus, there is a confluence of the pressures to create jobs and the existence of brownfield sites in the north and east, and the pressure not to develop in the south. This was thought to make the planners' role somewhat easier with a broad consensus existing over strategic direction for the county: 'We say to the green lobby we'll avoid green land and say to the development lobby they will have to recycle land and go where opportunities arise' (planner). KCC does, however, recognise that the differentiated nature of the demand for housing and employment land in Kent implies that it is not this simple. The irony is that those whom Kent wished to attract wanted to locate in areas that Kent found unacceptable. This is recognised in the County Council which promotes a portfolio of employment sites throughout the county to take advantage of this differentiated demand pattern, whilst a realistic view is taken of the future of Kent's more economically depressed areas.

The translation of environmental concern in development plans is spatially variable across the county, with districts traditionally concerned with protection of areas of high landscape value being a little more informed on environmental matters as a general rule. KCC has responded rhetorically, moving the environment chapter to the front of the deposit KSP, and subjecting the plan to a thorough environmental audit. Much of the debate has, however, been about accommodating development in a more sustainable way rather than working from an environmental capacity perspective. Thus it is not clear how the pressure for new employment opportunities and pressures on the environment are being mediated.

Plan-making processes

The politics surrounding the development of the CTRL are similar to those surrounding Channel Tunnel construction in the 1980s. As Holliday *et al.* (1991) observe, 'Kent qua Kent had no need of a Channel Tunnel, did not really want one, and could certainly not be made to benefit from one to the extent that Nord Pas de Calais could' (p. 109).

Development plans have played little part in framing decision-making regarding the CTRL. What CTRL construction, its associated road improvements, and the construction of a railway station at Ashford have led to is the potential realisation of land allocations contained in successive development plans. Ashford has been typically used 'as a sink [for allocations]' (planner), and the new international passenger station may finally kick start a sluggish market in this part of Kent.

The major strategic development opportunity in the county is Thames Gateway. The Kent Thameside initiative is an attempt to capture the benefits of this opportunity for north west Kent. In some senses it could be said to embrace New Right thinking on public–private partnerships. As collaboration between Blue Circle Properties (the current Chair), Dartford Borough Council, Gravesham Borough Council, KCC, and the University of Greenwich, the initiative seeks to take a thirty-year view of the area. Land-use planning is guided by RPG 9 (DoE 1994f), which has clearly been influenced by some partner organisations in Thameside. Market thinking has embraced the public sector, which strives to be seen working with the 'movers and shakers' in the area. To some extent the initiative operates outside the frameworks being developed in development plans, which is seen by some partners as allowing freedom to devise a vision for the area unconstrained by established planning policy. Others contend that an exercise on this scale will effectively steamroller the statutory planning framework when attempts are made to incorporate it in such a frame. KCC argues that partnership work in the non-statutory arenas bolsters the statutory planning work, helping the process of legitimisation, through explanation and maintenance of dialogue, especially crucial with big landowners. It remains, however, that each partner in Thameside 'trusts each other as far as they can throw them' (local authority employee), possibly reflecting the negotiations yet to be undertaken to extract maximum gain from the partnership and a history of difficult relations between many of the partner organisations, not least the county and the districts.

The Thameside initiative arose partly because the statutory planning system could not be made to work fast enough. Local plans were seen as being stuck in the time-warp of the mid-1980s when their preparation was started. The basic issues they addressed had remained the same despite the changed economic climate ten years later. We found that landowners and developers, often criticised for their short-term approaches to land holdings were going to planners and saying they'd like to discuss what's happening in Kent over the next thirty years but that planners in such circumstances 'fall off their perches' (property interest). They were not prepared to discuss matters beyond the time frame of the plan. 'They'll say you can't think about housing, it's not in the plan and we'll say "well no we want to think about the next plan or maybe the one after that" [but] they feel totally and utterly exposed because they have no policy framework on which to rely' and crucially 'someone might tell a Councillor and the planners will get it in the neck' (property interest).

Central vs. local: policy formulation

Issues of housing and infrastructure are two key areas for centre–local interaction in Kent. The SP (structure plan) is currently subject to a direction from the SoS to force Kent to provide for the housing numbers indicated in the RPG. The SoS cannot be seen to back down from RPG figures, with Kent being important in shifting development pressure from the west of London to the east. KCC for its part is working to ensure that its SP figures are translated into local plans by appearing at public local inquiries at the behest of the House Builders Federation to make sure districts implement DoE and county-level policy and allocations. However, the development industry remains concerned that some districts and the SP as a whole to an extent, are allocating land in areas that are unlikely to be attractive to the market. The LAs are therefore considered to be failing in their duty to translate central government household projections in a reasonable way to local level.

Against this local discretion, major decisions concerning the East Thames Corridor and the CTRL stations were made by central government with little local influence. This meant that development plans were waiting on decisions from central government and area strategy was hinging on government decisions over major infrastructure schemes. Development plans had to proceed whilst decisions were awaited, but it is clear that the decision to promote Ebbsfleet completely changes the land market in the area, and thus the entire planning context. Similarly at district level rapid shifts in central government transport policy had led to funding being withdrawn for road schemes in Maidstone and Swale. This has prejudiced land allocations in versions of the development plan in these areas.

Lancashire

The Lancashire SP, *Greening the Red Rose County*, is at a similar stage to the Kent SP. An Examination in Public (EIP) has taken place, the inspector's report has been received, and Lancashire County Council (LCC) is preparing modifications at the time of writing. The SP updates that adopted in 1990, but is intended to provide guidance only until 2006. The deposit SP has eight aims. The first of these is 'to make significant steps towards sustainable development and growth' (LCC 1994). It was however clear from our interviews that despite a great commitment toward sustainability issues the key issue concerning the County Council was that of promoting job opportunities.

Plan driving forces

It is clear that the central driving force behind LCC policy is that of creating employment opportunities. Concern for the environment arose separately, out of issues of water quality originally, and links are now actively being made around quality-of-life issues. Considerable attempts have been made to integrate these environmental concerns into the SP, including a detailed environmental audit of the previous SP and the draft version of *Greening the Red Rose County*. The feeling amongst some stakeholders we interviewed was that a booming economy is needed before you can talk of green issues. This seems to be based on an untested presumption that local economies are going to improve. This presumption does not exist in Kent for two very different reasons. First, an in-depth consideration of some local economies has led to the conclusion that whilst considerable efforts will be made to create job opportunities in these areas, other alternatives need to be looked at including the management of gradual economic decline. Second, there existed a strong feeling that mid-Kent was in effect 'full' and should be protected from development.

Most stakeholders in Lancashire felt that although Lancashire's environmental work was admirable, 'we see that greening, environmentalism, is seen as a component [by the County Council] rather than as an all embracing issue and it can be over-ridden by other components . . . the most important of which is economic development' [pressure group]. This was a criticism by many, but in some senses it should be seen as an exertion of local political choice by the County Council.

The introduction of sustainable development objectives in the structure plan has led to conflict with other objectives and policies such as the commitment to a large number of road projects in the plan. This conflict is compounded by roads being essentially traded as political goods, with members refusing to drop pet schemes that they and engineers have invested in. This continuing desire for road-building had brought LCC into conflict with the Government Office for the North West (GONW) who objected to a number of schemes in the SP.

Plan-making processes

Lancashire has played a large part in the North West region's collaborative attempts at producing a broad spatial strategy to attract investment, and to demonstrate a co-ordinated approach to EU funders and investors alike. In many ways the SP reflects this strategy, building as it does on the concept of corridors for investment, and strategic development locations. Throughout the plan the environment is treated as a backdrop for investment and place marketing. A significant role though has been played by county planners to drive forward the importance of other environmental considerations, through specifying robust policies on pollution and landscape conservation. In each case their efforts have been diluted by the combined action of GONW and other interests within the County Council itself.

The EIP provided an important arena to exchange ideas, make deals, and progress policy on a number of issues. Chief among these were: a site for large-scale business development; road schemes; and trading between the districts on housing allocations. There was very little public involvement in the production of the SP but it was seen as a 'reasonable method for getting consensus between those with a statutory responsibility' (planner).

Central vs. local policy formulation

Inter-governmental tensions are rife in development plan production. Nowhere was this more evident than in housing land allocations. Some authorities felt it was up to them to determine housing numbers locally: 'the Government Office and the centralisation of housing numbers is quite rightly seen as a power threat . . . [and] the RPG is very wrong in settling the figures to come to strategic authorities as a bit of technocracy rather than politics' (planner).

Our interviews also showed, however, that the right of last reply which local authorities have in policy formulation is being questioned by other participants. The self-adoption of plans was seen as 'totally flawed' by one property interest. The only way to change the policy direction of a plan was to become involved in the build-up to its production, which involves more work for developers and LPAs. Sites were seen as more difficult to get than ten years previously, and with the success rate on appeals reduced to about 25 per cent, this avenue was no longer seen as important. In terms of development opportunities, LAs were not prepared to consider things beyond the timescale of the plan. Government Offices were not seen as a positive influence by either developers or LA planners, who felt they were totally marginalised on the issues that matter and were not prepared to interfere anyway. Local authorities were thus regarded as 'tak[ing] no bloody notice whatsoever. They follow their own agenda, and government guidelines are deliberately vague enough to allow them to do that' (property interest). Certain stakeholders, particularly in the development industry, would support a call for inspectors' recommendations to be binding on LAs.

Shifts in central government transport policy have affected development plan strategies. Lancaster and Chorley districts found themselves in the difficult position of being unable to progress housing allocations in plans as a result of withdrawal of funding for road schemes on which development plan strategies' depended.

West Midlands

A discussion of New Right attitudes and planning in the former West Midlands County must start with the abolition of the County Council itself. We have explored how the metropolitan districts are coping without a strategic institution at the level of the conurbation. We also investigate how this and a fragmentation of service delivery in other governance aspects, are affecting urban region management.

We see a great deal of continuity in policy dating back at least to the 1982 West Midlands SP (WMSP). This continuity reflects this history of collaborative working and the store of ideas that has built up. The WMSP set the context of an urban regeneration focus for the conurbation which has since been supported in the region as a whole, and through the strategies and UDPs prepared in the conurbation.

Plan driving forces

Environmental issues are less prominent in the West Midlands UDPs than in the plans of most of the districts we looked at, and certainly less so than in the structure plans of Lancashire and Kent. This is partly the result of the time of UDP production being somewhat before a broader definition of sustainability entered the mainstream of UK plans. It is also a reflection of the lack of political currency such issues have in the conurbation.

What the plans do reflect, however, is a concern for local economies and social issues. The regeneration of the conurbation is a common theme with all interests united in their aim to stem the loss of population and employment from the urban area. Early drafts of Birmingham's UDP also reflect a little more the boosterist strategy the authority pursued in the 1980s. This led to substantial pressure from some stakeholders for change, notably around two issues: the redevelopment of the City Centre, and the South Birmingham M40 link road proposals.

Plan-making processes

In the metropolitan area a strategic decision was made by all the metropolitan districts to prepare their UDPs as quickly and painlessly as possible. Certainly in terms of speed of preparation this has been achieved with six out of the seven districts having had an adopted plan in operation since 1993. This speed of preparation was achieved largely by rolling forward existing plans but was critically dependent on the capacity developed at conurbation level prior to this. This essentially consisted of a history of working together, and a consensus over policy direction despite tensions between the districts over some issues. This provides a store of ideas and intellectual and social capital to draw on when dealing with strategic issues.

The centralisation of policy formulation

The creation of two Urban Development Corporations (UDCs) in the West Midlands has diminished the power of the development plan and the local authorities. We found considerable dissatisfaction amongst urban region stakeholders that decisions taken by UDCs, primarily by the Black Country Development Corporation, were undermining adopted policies in the UDP. This was particularly pertinent in the context of out-of-centre retailing and policies aimed at maintaining existing district and town-centre viability. Also, employment land being released for retail has been thought to undermine some of the technical inputs to UDP production. This situation reflects a conflict in issue agendas. The UDCs are under pressure to maximise revenue prior to their dissolution, whilst LPAs are concerned with maximising the legitimacy of the plan and with satisfying member, trader and environmental lobbies focusing on development in existing retail centres.

There has also been some disquiet over recent processes aimed at allocating large employment sites for inward investors, not least because the process is felt to be taking place outside the planning system. It is felt that when sites are chosen by a group of selected interests, including the Government Office for the West Midlands (GOWM) they will enter the planning system for debate but with considerable advocacy and weight. The GOWM has been involved in the selection process and will be judge and jury on the report's conclusions. Further criticisms of GOWM's role came over its insistence on pursuing the housing numbers game without paying significant attention to the issue of deficient effective demand. Local authority attempts to highlight the problem of vacancy rates in the conurbation and to use this as an argument to restrict greenfield development has not been accepted by GOWM and development plans have had to reflect this.

Case-study discussion

Plan driving forces

A central feature of our work was to examine the emergence of an environmental agenda in plans. It was clear that in Kent and Lancashire environmentalism had great political currency, coupled with grass-roots support in Kent, and this had filtered into the production and content of plans. The West Midlands plans are less concerned with environmental issues but are a product of their time. Planners in the conurbation saw them as conforming to the emergent environmental agenda in any case. In other areas planners were expressing concern that they would like to approach the plan from the starting point of environmental capacity but had no idea how to go about it or whether it was possible.

It remains, however, that the UK, in comparison to other European countries, is weak in terms of pursuing a green approach (Marshall 1994, 1994a). Marshall argues that the diversity of actors, and the nature of local political economic contexts, make a shift toward a more radical green agenda unlikely. The result is a policy manifestation where broad environmentalist statements are explicit but are placed second when issues of economic growth or people's lifestyles are concerned (Marshall 1994). LAs are, perhaps wrongly, fearful that a highly regulative approach with regard to environmental standards may disadvantage the locality in competing for investment as well as incurring the wrath of existing local business interests.

Part of the explanation for this may lie in the local political currency of environmental and economic issues. Our three case-study areas offered different local conceptions of the 'environment' and potential threats. In Kent there was considerable grass-roots support for environmental issues and this fed through members and into policy. In Lancashire environmentalism entered the political agenda from politicians themselves who were looking for 'issues', despite the presence of vocal lobby groups in the county prior to this. In the West Midlands interest seems to be officer- and partly member-driven with little grass-roots pressure. This is partly historical. Kent has a long tradition of physical environmental protection and associated lobbies. This is mirrored in parts of the other two areas, but at a more general level issues of investment predominate.

Plan-making processes

Even if the additional legal power ascribed to the development plan through S54A is coming to be regarded as minimal (Herbert-Young 1995; MacGregor and Ross 1995), the fact remains that local authority officers, business, developers, interest groups and the general public perceive a change and are acting accordingly. As a result of the well-known subsequent pressure on the system, notably the inquiry process, the resource costs to a local authority can be enormous. In addition the cost to the legitimacy of the plan when it takes a great many years to get to adoption is substantial. The solution might be to abolish the number of plan stages. This is possible but our experience of areas where this happens shows that they were only able to move quickly to adoption due to the plan-making capacity built up through previous work with a range of outside interests. The relation-building work with local stakeholders and building up agreement over issue direction had taken place previously. Thus in Ashford considerable work on building institutional capacity through large-scale consultation had given the council a mandate to go straight to this stage without a deleterious effect on local democratic concerns.

Key players in development plans do not appear to have changed in recent times. Business is unhappy about many aspects of the planning system and about individual proposals. Such interests are being told to 'get in the inquiry with every one else' (civil servant). This represents a clear shift from the insider status some groups had in the 1980s (Holliday 1993). This appears to be a change from the Thatcher administrations in that certain privileged groups appeared to operate behind the scenes and get what they wanted rather more directly. The roads and housebuilding lobbies could be considered to be two of these (Holliday 1993). They are now being denied this privileged insider track. Thus we see a visible change under Major in that Thatcher was noted for her dislike of consultation, and consensus politics more generally. This may reflect Major's concern for a more equitable approach or a reaction to increasing policy overload in Whitehall, or possibly the effects of two key players (environmental and development lobbies) providing counterbalancing forces.

A consequence of the 'new' importance ascribed to the development plan appears to be that local authorities are tending to hide behind it and feel unable to operate beyond it. Thus, landowners and developers who wish to take a long-term view of their land banks are sometimes thwarted by LPAs' failure to consider anything beyond the lifetime of the plan and what it contained. Plan visions are being seen as too short-term by larger landowners who are looking for more market certainty and wish to take a considered view of their land holdings. The 'plan-led system has led some planners in some authorities to retreat behind the plan, its policy and its timescale' (property interest). New Right attempts at a greater 'market-orientation' for plans have therefore been a mixed success. Whilst LAs pay possibly more attention to economic concerns, the idea that plans are flexible and responsive to market demand appears not to be borne out. The requirement to prepare district plans and the introduction of section 54a, coupled with increasing environmental emphasis on restraint, appear to have thwarted attempts for plans to achieve these aims. This may partly be a fault of the district-wide plan idea. Suttie (1994) observed that in the preparation of a local plan for Banff and Buchan there was a noticeable negative correlation between settlement size and the level of interest shown by the public. The old system of action area plans (provides a more local focus), and subject-specific (easier to identify stakeholders), would appear to satisfy these requirements in a way that district-wide plans do not.

There is developer pressure to move from McAuslan's 'administrative nexus' to a market-driven nexus backed up by legal rights (McAuslan 1980). Developers are becoming impatient and averse to local political discretion. This could be seen as purely an endogenous change in our case-study areas, but is more probably a build-up of entrepreneurial frustration unleashed by the ideological hope of successive Conservative governments. The role of the local authority in spatial change has become one of resolving competing claims over land use (Adams 1994). Their dependence on developers and local businesses to effect change suggests they will actively seek these implementors as partners or consultees in plan-making (Rydin 1993). However, local political concern over loss of greenfield land in particular are highlighting tensions between the economy and the environment.

Central versus local control

The degree of central control over local planning policy content has increased under Major's leadership as a result of three factors. First, the increased use of PPGs, which have taken on a wider remit than circulars, and provide a fuller framework in terms of agenda setting. Second, greater DoE specification of the procedures of plan-making. Third, an increased procedural role for the involvement of market-oriented consultees. The increase in guidance from central government has reduced the possibilities for local diversity in plan approach and expression. This is not necessarily central capture of local debate, but through the requirement that (major) development proposals should emerge through the development plan process, it ensures that developer interests may come to dominate the eventual policy direction. Local discretion has, on this premise, to secure substantial local support from significant interests to survive the long process of policy formulation.

This was obviously not the perception of those developers we interviewed and who commented on LA control of the plan process. This group of respondents felt that the move to plan self-adoption gave an LA *carte blanche* to do what it wished, and in particular to ignore the inspector's report. They argued that the overriding importance of S54A, which was likely to reduce the chances of planning permission by appeal, appeared to swing the pendulum back towards local political discretion. It is recognised that developer interests do not necessarily coincide and, whatever the plan policy, some will be disadvantaged, but some clearly felt that highly organised NIMBY groups were distorting otherwise open technical processes of land allocation. These feelings may be given added currency because of the diminishing influence of some of these powerful groups at the centre.

Partnership working was firmly on planners' agendas in all areas. This raises interesting questions for the planning system. Such processes are involving an increasing number of stakeholders although some local authorities were thought to retain too much control over participation in some cases. Partnerships do, however, arguably give LAs more power through the leverage of resources for mutually beneficial objectives. In addition most partnership work takes place outside the planning system. There are fears that such corporatist arrangements enter the planning system merely to legitimise decisions and when they enter the system they do so with great advocacy.

In some issue areas, plan policy is a land-use translation of a centralised process. This is particularly so in transport and minerals, and in housing land allocation. Transport policy, because of the fiscal and regulatory dependence of the local on the central, has to read and conform to current Department of Transport thinking to get things done. Cuts in roads expenditure initiated by central government have upset land allocations and indeed entire planning strategies in parts of our case-study areas where plans hinged on road schemes which five years before had looked highly likely to get funded.

In addition the removal of some areas from local authority jurisdiction has created tensions. The creation of UDCs in the West Midlands has clearly led to decisions which many stakeholders felt did not conform to national, regional and particularly often, local planning policies. The short-term, receipts-driven approach of UDCs was felt to be undermining their planning credentials. Similar points had been made in reference to the approach of the Commission for New Towns.

Conclusion

During the 1980s planning doctrines were watered down to resurface by the end of the decade in the context of new debates. The Rio accord and EU Directives have questioned the British sectoral approach to spatial planning considerations. This has helped to salvage planning rationales for intervention in market decisions. New issues have come to frame agenda setting in the 1990s:

1 Treasury funding commitments and cutbacks.
2 The translation of sustainability arguments into PPGs. This has impacted on the spatialisation of housing which in turn has added weight to Green Belt designations. There is growing agreement and understanding of environmental principles but little consensus on the most effective means of reducing environmental pollution. Some stakeholders have called for a complete rethink about what a sustainable approach might be and whether for example Green Belt policy is detrimental.

Both issues have impacted on spatial policy, questioning the need for new infrastructures and bringing about a more holistic approach to mobility and environmental assessment generally, onto the policy agenda. Both the constraints on public-sector expenditure and the time lag in implementing decisions already made, in each of our case-study areas, prevents the full import of new thinking being delivered.

Thatcher brought new private-sector values into the policy implementation arena. Major's contribution has been to 'democratise' decision-making through ensuring those development decisions which will have significant land-use implications are legitimised through the local plan mechanism. Nationally significant decisions, as always, are protected from this proto-pluralist debate. Local stakeholders, developers and local authorities, must now market their proposals and secure local support through development plan arenas. Development plans, as a result, are guided by an overlapping and complex structure of agencies, the management of whose interaction is now on the policy agenda with the trend again to longer and costlier plan-making processes. Thus, the New Right has not achieved deregulation but merely negotiated the balance of regulation in different places.

Central government policy criteria have effected changes not only in the content but also in the procedures of policy formulation in the 1990s. PPGs have ensured a level of consistency through providing a framework for interaction. However, PPGs coupled with decision monitoring have the consequence that real local choice cannot be exercised. This illusion or denial of choice could be seen as an attack on the role of the local authority planner in identifying community needs, or it could be interpreted as an admission that handing real choice to the local level is undesirable either for political reasons, or for the protection of interests beyond that of the local. That is, too much local discretion could hamper wider political objectives of planning or other regulatory systems. Local political problems encountered in accommodating the government's revised household projections in terms of housing land allocations provide good examples of this throughout much of England and Wales.

Because of the level of technical skills and negotiation required, access to decision-making arenas has been dominated by project-led development companies. Their resources enable them to participate in these arenas and they are increasingly courted by cash-starved local authorities, who need private-sector ideas and matching funds to secure government funding. Local plans are effectively sidelined in local authority bureaucracies because of their inability to consider economic and social issues beyond land-use considerations. Bruton and Nicholson (1987) suggest this separation of spending plans and policies of local authorities and other public agencies creates tensions between the public and those with a governance role. This is compounded by the level of activity in other governance arenas: notably attempts to (a) capture government funding – exacerbated by the competitive approach to funding local authorities – and (b) bring in private-sector investment. The land-use plan, which cannot respond quickly enough to such initiatives, has a less prominent role in the culture and work of the local authority. West Midlands LPAs which tried to make plans flexible enough that short-term financing could not affect the thrust of the plan, were almost thwarted in their approach by central government civil servants wanting to see greater certainty in plan proposals.

We have a planning administration that is moving towards more precise rules and criteria, more substantive specification, along what might be termed a more European route. The question arises that if policies are well devised, do they need to be specifically spatially articulated? Would a preconceived general framework such as employed by the government in its UDC, SPZ and EZ experiments serve to reduce bureaucratic delays and encourage investment? As one of our interviewees remarked, a core shopping policy or a policy controlling development in an Area of Outstanding Natural Beauty is the same in one part of the country as another; 350 planning authorities reinventing the wheel is not necessarily healthy. 'It's professional mystique, isn't it, that we are all doing something unique in individual contexts' (district planner). If the ideological preference of central government was for an extension of the 'preconceived general framework' concept in planning, how would this change the purpose of town planning? Or could a renewed emphasis on the spatial provide a means to break the centrist British state?

A further shift toward support for the market would have a deleterious effect on public participation in producing local land-use strategies, through reduced political involvement. Development proposals would be seen as *faits accomplis* with appeal only through the SoS. An alternative view lies in a collaborative approach at local level with more discretion for LAs to determine how their areas are shaped. This approach may require a broadening of the remit of the planning system, possibly using sustainability or quality-of-life notions as guiding principles. These two directions mask a range of tensions for a new left government to grapple with, just as such tensions exposed the deficiencies in New Right thinking in the late 1980s.

Acknowledgements

ESRC Grant Number R000235745, Development Plans and the Regulatory Form of the Planning System. The authors wish to acknowledge the input of the other team members, Patsy Healey and Tim Shaw at the University of Newcastle and Simin Davoudi (University College London).

6

SIMPLIFIED PLANNING ZONES

Philip Allmendinger

Introduction

Simplified Planning Zones (SPZs) are one of the 'great unknowns' of the New Right. References to them in planning texts fall into one of two distinct categories: perfunctory descriptions that pay lip service to their (rare) existence (e.g. Rydin 1993; Cullingworth and Nadin 1994) or alarmist visions that conjure up an ideologically driven attack upon planning (Thornley 1993). This neatly sums up the confusion surrounding SPZs and planning as a whole during the 1980s and 1990s. It is too much to characterise them as a symbol of an era (after all, there are only half a dozen or so in existence) but neither should they be dismissed as a failure. In reality, SPZs mark the high and low points of the Thatcherite approach to planning – high in terms of their wide deregulatory intentions, low in terms of their lack of practical effect and use. Ideologically and in the purest form they represented a clear departure from the whole post-war approach that would have revolutionised planning; they combined plan and permission, effectively removing the discretion at the heart of the UK system. However, by the time the first zone was adopted in Derby in 1988 they resembled little more than glorified development briefs. And for an initiative that could have feasibly covered just about any area of the country (excluding some environmentally sensitive areas after successful lobbying by shire Conservative MPs) only a dozen or so were ever attempted. Mrs Thatcher was famed for her strident and often hectoring approach summed up in the acronym TINA: 'There Is No Alternative'. How did a distinctly Thatcherite approach to planning become so utterly altered? As the approach to this book suggests, the answer lies in a combination of central approach and local use of SPZs.

The origin and evolution of Simplified Planning Zones

Unlike Enterprise Zones, Simplified Planning Zones have not benefited from much academic interest or research. Most of the work has been speculative (Lloyd 1987; Rowan-Robinson and Lloyd 1986) or derived from comparisons with Enterprise Zones (Job 1984; Thornley 1993). Recent work by Cameron-Blackhall (1993) has briefly reviewed the concept and take up of SPZs and postulates on the experience so far. Apart from Department of the Environment sponsored research by Ove Arup on the adoption procedures following criticism that they were too lengthy (DoE 1991b) there is little empirical work. Most of what is known about SPZs comes from monitoring reports published by the individual zone authorities.

The origin of Simplified Planning Zones can be found in the simplified planning regimes of Enterprise Zones. Most commentators agree that Enterprise Zones have their lineage in the ideas of Peter Hall who argued that inner-city industrial decline was not natural or an inherent feature of advanced capitalist economies but a consequence of state intervention – particularly planning (Banham *et al.* 1969). Hall developed this idea in his most influential 'non-plan' speech in 1977 when he claimed that urban renewal would require 'highly unorthodox' methods: in essence a 'non-plan' where selected areas of inner cities would be exempt from planning and other controls (Hall 1977). The Conservative opposition of the time fell on Hall's ideas and applying the rhetoric that characterised their alternatives renamed Freeports as Enterprise Zones (EZs). The Shadow Chancellor, Geoffrey Howe, wasted little time in implementing the idea once in office and announced in March 1980 that the government would approach six or seven local authorities and invite them to submit proposals for sites of around 500 acres. Following negotiations with local authorities twelve zones were created in the first round of EZs in 1981, a further fourteen in 1983 and one in 1993. Apart from the subsidies available the zones also contained a 'simplified planning regime' that combined plan and permission for a range of uses and buildings.

Assessment of EZs was undertaken in government-commissioned reports for the first three years by Roger Tym and Partners (1982, 1983, 1984) and later in annual Department of the Environment (DoE) published data. The evidence presented is basically a catalogue of change that does not reveal the mechanisms by which that change has occurred. Almost everyone – planners, developers, occupiers – agreed that the relaxed planning regime had had virtually no effect on the kind of development that takes place (Hall 1984) and the Tym reports also conclude that very little of what has been built was in conflict with planning policies previously in force. The reasons for this general compliance, the report concludes, is that most of the EZ areas were more suitable for industry or warehousing and were allocated for such in the erstwhile development plan. The zone merely accelerated development (Roger Tym and Partners 1984: 118). The monitoring reports also conclude that there was no deterioration in design standards and that landlord control and building regulations were a major influence on this. A sample survey of developers sought views and experience of the new planning regime. Only one respondent considered that the simplified planning regime was a 'critical' factor in reaching his/her decision to develop, two said it was 'important', eight regarded it as of 'minor importance' and ten 'irrelevant'.

Though the benefits of the simplified regime in EZs was questionable the DoE specifically asked Roger Tym and Partners to look into the possibility of extending them before their final report (Job 1984). In June 1983, a year before the final report was published, the DoE circulated a paper for ministers that considered extending the schemes which concluded that

> the principle of defining an area within which the limits of planning control . . . are defined, and removing conventional discretionary planning control over the types of development, offers a very interesting alternative to the system that has existed since 1947. The area approach offers the possibility both of testing the simplified system and also perhaps of changing attitudes towards the proper purpose and extent of development control.
>
> (DoE 1983)

It was clear from these sentiments that the government considered SPZs would form part of a wider examination of the post-war planning system and this was confirmed by the then Secretary of State, Patrick Jenkin (1984).

Although the final Tym Report (Roger Tym and Partners 1984) was less than enthusiastic about extending the simplified planning regimes of EZs the government proceeded to issue a consultation paper on 11 May – shortly after the final Tym report. The central premise of the paper (which for the first time used the phrase Simplified Planning Zone) was that conventional controls over property and development inhibited private-sector investment and enterprise:

> Instead of subjecting all development proposals to the uncertainty and delay of discretionary planning control, the SPZ schema would specify types of development (including specified categories of outdoor advertisements), allowed in the zone and the conditions and limitations attached:
>
> (DoE 1984d: 2)

The planning system was painted as reactive, negative and time-consuming while SPZs would offer speed and certainty as well as allowing local authorities to pursue more positive approaches than is possible with traditional development control (DoE 1984d: 2). The link between EZs and SPZs was portrayed as the latter building upon and extending the success of the former. Although SPZs would not be appropriate in National Parks, Areas of Outstanding Natural Beauty or Conservation Areas they would be suitable for any other area.

Responses to the White Paper from a wide variety of companies, voluntary bodies and government agencies did little to encourage the government to proceed. Virtually all of the consultees, including private-sector interests, believed that SPZs would not achieve the government's aims. Most, including the Welsh Office, rejected outright the notion that planning inhibited development by creating uncertainty and delay. Similarly, there was widespread criticism that SPZs built upon the 'successful' experiment of EZs. According to the Association of District Councils (1984) the link had 'bred suspicion' of the government's intentions.

As well as the principle and background to the zones, concern was also expressed at the detailed provisions, especially the wide scope for SPZs to be adopted in a variety of areas. Bodies such as the Campaign for the Protection of Rural England, the Nature Conservancy Council and the Countryside Commission all pressed the government to further restrict the scope of SPZs by using an inclusive rather than exclusive designation system similar to EZs. In a letter to the DoE summarising a meeting between them, Slough Estates (a major private landowner involved in the preparation of SPZs) agreed with civil servants' views that in practice SPZs would be used only in certain circumstances which would tend to be urban areas in need of regeneration. This began to cast doubts over the locational characteristics of SPZs; officially they were deregulatory though unofficially they were seen as selective. Similar confusion surrounded the relationship between SPZs and development plans. The government had not clarified this and a number of bodies including the Association of Metropolitan Authorities and the Association of District Councils felt that SPZs should conform to development plans. The role of the Secretary of State was also debated with developers and land interests wanting intervention if local authorities tried to impose unnecessary conditions on the zones while bodies including the Royal Town Planning Institute (RTPI) questioned any further centralisation of powers: 'It is perverse that a proposal allegedly designed to "reduce red tape" and bureaucracy should provide such major intervention opportunities for central government' (RTPI 1984).

The revised consultation paper that formed the basis of the 1986 Housing and Planning Bill introduced some important differences including:

1 no upper or lower limit on the size of the zones
2 any person could request a local planning authority to adopt a zone and appeal to the Secretary of State if refused
3 ministerial powers to intervene and alter the zone that is being prepared or adopted or to direct a local authority to adopt a zone

4 a reduction in consultation requirements
5 a ten-year life-span for the zones.

These changes saw a definite shift towards the concerns expressed by production interests. The critical rhetoric that accompanied the 1984 paper was only slightly toned down in the 1986 Bill though Conservative backbenchers used the opportunity to rail against local government and the planning system. During one debate on the Bill Richard Tracey, a minister at the Department of the Environment thought it was 'fair to say' that the vast majority of local planning authorities were 'sensible', though sometimes slow, though the Conservative Cecil Franks saw the Bill itself as an acceptance by the government of their unreasonableness: 'the raison d'etre of the Bill . . . is the failure of local bureaucracy . . . we are talking about the failure of planners' (Hansard, 13 March 1986, Col. 328).

Although these assertions were questioned by some opposition MPs others pointed to what they saw as the real inhibitions to development: high rents, VAT, rates and the government's deflationary policies (D. Thomas, Hansard, 4 February 1986, Col. 214). The government's desire to concentrate more on the positive aspects of the zones and play down any deregulatory content and criticism of the planning system led to some confusion among MPs. Ministers were aware of the lobbying by many conservation bodies which had been used by some government and opposition MPs to portray SPZs as a 'planning free-for-all', an image that in shire areas was not electorally popular (see Bishop's chapter in this volume). Similar confusion arose over the locational characteristics. The Bill had left the question ambiguous and the government stressed the importance of being 'flexible' and not tying the zones to any particular purpose or area. However, MPs from all parties (though in particular Conservatives) left the government in no doubt that SPZs were more suited for urban areas: 'I must make it clear that in putting forward the amendment my honourable friends and I were not proposing a Conservative [sic], backward looking measure. We seek to protect the environment of the south of England' (Hansard, 24 April 1986, Col. 539).

The government consistently resisted such pressure (though it consistently accepted that SPZs were more *likely* to be found in urban areas): 'We have very little idea of where SPZs are to be. Local authorities are to decide where and how these powers shall be used' (Allen Stewart, Parliamentary Under Secretary of State for Scotland, Hansard, 13 March 1986, Col. 342).

But it was this discretion for local authorities that worried some Tory MPs who felt that local authorities were hostile to development and that the power of designation should lie with the Secretary of State. The government rejected this view and instead pointed to the Secretary of State's reserve powers and the widespread consultation that would accompany any zone. Some MPs wanted reassurances that the zones would not be used to overcome government or national policies by pernicious local authorities. Again ministers resisted: 'I shall not list the kind of opportunities which local planning authorities ought to be more than capable of deciding for themselves' (John Patten, Hansard, 18 March 1986, Col. 348).

If the government would not fetter the use of the zones and reassure MPs, surely they would make sure that unsympathetic councils could not impose conditions that ran against the spirit of the schemes? '[Local authorities] will know best what developments to allow and what conditions, if any, are necessary' (Richard Tracey, Hansard, 13 March 1986, Col. 339).

The government's reluctance to specify any aims or objectives for SPZs led one opposition MP to conclude that they were not quite the mad dash for freedom that some government members had wished for (John Cartwright, Hansard, 13 March 1986, Col. 343).

There seems to be three reasons for this. First, from the responses to the 1984 consultation paper it was clear that some bodies, especially conservation groups, felt that deregulation of the planning system in this way was a step too far. Some of these bodies, as Hirst (1989) and Elson (1986) have pointed out, are the 'natural' supporters of the Conservative Party and the government needed to rethink its approach so as not to alienate this support. Second (though linked to the first reason), the government's insistence that SPZs were building on the 'success' of Enterprise Zones both alienated some people and implied that the zones were a form of urban regeneration. Although the government recognised that the zones were more likely to be used in urban areas they did not want to limit the possibility of their being used elsewhere. Additionally, the government also wanted to play up the 'positive' aspects of the zones in order to move away from the 'urban dereliction' image that followed EZs. Finally, the Tym reports on EZs had provided both government and opposition MPs with arguments to support their views. The government realised that providing SPZs with aims like EZs' would mean that they could be evaluated. As the simplified planning regimes of EZs were the basis of SPZs and had been shown to be the least effective part they sought to remove any scope for monitoring.

The result of this was a vacuum where the objectives of SPZs should have been. As implementation was to be left to local agencies this allowed the substitution of other aims for the zones.

The use and impact of four Simplified Planning Zones

To answer the question set at the beginning of this chapter, i.e. how did a distinctly Thatcherite approach to planning become so utterly altered, we need to examine how the zones have been used and what their impact has been. In all there are seven adopted zones in the country, and a further six in various stages of adoption (though some of these such as Cleethorpes and Birmingham have effectively been abandoned).

I examined four zones – Birmingham, Cleethorpes, Slough and Derby – following what Hakim (1987) has termed 'deviant' case-study design which requires certain criteria to be identified that enable comparisons to be drawn between the ideal and the experience (i.e. what the government intended and what actually happened). Examining the twelve zones that have been attempted or adopted against a number of government aims and characteristics it was possible to identify clear deviants. Four criteria were used (and are summarised in table 6.1):

Use classes

PPG 5 (Department of the Environment 1988) states that SPZs are more suitable for homogeneous areas where proposed use classes will be roughly the same thereby limiting the number of conditions. Whilst most zones had on average three permitted uses, Birmingham, Cleethorpes and Slough had either five or six.

Aims and objectives

The government's aims for SPZs can be seen in PPG 5 and can be summed up as providing certainty and flexibility in the development process thereby promoting development or redevelopment. While all the zones

Table 6.1 Selected data on Simplified Planning Zones

Zone	*Status*	*Permitted use class*	*Number of sub-zones*	*Number of conditions*
Corby	Adopted	B1, B2, B8	1	8
Derby	Adopted	B1, B2, B8	2	16
Grangemouth	Adopted	B1, B2, B8	5	12
Highlands x2	Adopted	B1, B2	1	13
Birmingham	Proposed	B1, B2, B8, A1, A2	7	36
Cleethorpes	Proposed	A1, A2, A3, C1, C2, C3	2	14
Delyn	Proposed	B1, B2, B8	3	14
Enfield	Proposed	B1, B2, B8	1	18
Monklands	Adopted	B1, B2, B8	1	8
Rotherham	Proposed	B1, B2, B8	3	14
Scunthorpe	Proposed	B1, B2, B8	4	21
Slough	Adopted	B1, B2, B8, A1, A2, A3	8	9

paid lip service to these objectives a different picture also emerged. Both Derby and Corby raced to adopt the first zone for promotional purposes. Slough Estates (the single landowner involved in the Slough zone) had been closely involved in the formulation of SPZs and was therefore very keen to pursue one. Cleethorpes was suffering from political instability and inconsistency in decision-making and looked to an SPZ to guide members in making decisions. Birmingham City Council was looking for ways to fend off an Urban Development Corporation, fearing the loss of control that would entail. As a condition of supporting an alternative private-sector-led initiative, Birmingham Heartlands Ltd, the Department of the Environment required the council to adopt an SPZ.

Conditions and sub-zones

PPG 5 states that sub-zones and conditions should be kept to a minimum. Table 6.1 shows how certain zones seemed to have up to three or four times the number of conditions and eight times the number of sub-zones of others.

The government consider three factors to be important in the use of any zone and these provided the criteria against which to compare ideal and experience:

1. Physical Suitability: the size and character of SPZs can be varied to suit different objectives and prevailing local circumstances including land allocated in a development plan, large old industrial areas, new employment areas, large-ownership sites, new residential areas and inner-city housing areas.
2. Preconditions: the three main criteria that the government feel should precede an SPZ are uncertainty, inflexibility and delay in the planning system.
3. Aims: the aims for SPZs include the promotion of development or redevelopment, speed of decision-making, certainty and flexibility in the development process.

Each of the four zones chosen, Birmingham, Slough, Derby and Cleethorpes, was examined against these criteria.

The Birmingham zone

Birmingham, with its population of around 1 million dominates the West Midlands region and, in line with other UK cities, it has suffered population decline since the early 1970s (Champion and Townsend 1991) and overall employment decline. The Saltley area which lies within the 'Birmingham Heartlands' was once the industrial centre of the city. The area has endured acute economic restructuring and accompanying social problems and is characterised by low-density industrial uses, vacant sites, derelict factories, and poor housing and social conditions. Since 1978 one-third of all manufacturing jobs in the area have been lost while unemployment rose from 6 per cent in 1979 to 15 per cent in 1985 – the highest increase of any region in the UK (Smith 1989).

In response to this the City Council has sought to diversify its economic base following what Loftman and Nevin (1992) term the 'prestige model'. This approach involves promoting a new national and international role for the city and in particular the centre. Such an approach involved a pragmatic view of the government's attempts to impose private-sector development-led urban regeneration strategies. As Loftman and Nevin point out, this pragmatism was in part due to vacillating political control and a tradition of cross-party support for major initiatives which enabled the city to avoid an Enterprise Zone. Urban Development Corporation (UDC), Housing Action Trust and, as we shall see, a Simplified Planning Zone.

It was in November 1986 that Birmingham City Council seriously began to examine the practicalities of a co-ordinated regeneration of Birmingham Heartlands and set up an Urban Development Agency (UDA). The UDA would aim to channel money from existing budgets to implement a regeneration strategy for the area. Agreement was reached with five volume housebuilders to redevelop large parts of Heartlands for housing and light industrial uses. By mid-1987 the government was looking for possible sites for the second round of UDCs and Birmingham was an obvious choice. Both the City Council and the developers realised that this could lead to a loss of control so a delegation from both successfully persuaded the DoE that the existing public–private initiative was working well. At this time the DoE was also promoting SPZs and linked the two together – a UDC would not be imposed if Birmingham adopted an SPZ. This was accepted, and Saltley was chosen by the Council as it was relatively 'self-contained'.

Early discussions with the DoE concerning the zone revealed widely different interpretations of the zone's function: the DoE was pushing for a wide range of uses while different departments within the Council objected to any deregulation. In the end the DoE got its wide range of uses and the Council limited any impact by the widespread use of conditions.

This exposed the paradox of SPZs. The wider the uses permitted (i.e. greater deregulation) the greater was the potential for conflicts between these uses and the greater the need for conditions to limit this conflict.

Birmingham Heartlands Limited (BHL) (the public–private company set up to redevelop the area) had had some degree of success in the redevelopment of Heartlands which led other developers to begin to buy sites in the area. The outcome was a significant rise in land prices which limited the ability of BHL to carry out some of the proposed works. All development halted, however, with the onset of the property recession in late 1989. The Council then came to the conclusion that having achieved this success it needed another source of funding to carry on with its strategy. Both officers and members agreed that providing satisfactory representation could be achieved on the board of any UDC they would not be averse to the idea of a Development Corporation. A deal was therefore struck with the DoE which secured a (unique) 50 per cent Council representation on the UDC board. There was also agreement that the City Council would determine planning application for the corporation and follow existing public consultation procedures.

With the UDC came the decision that the SPZ was no longer required and it was consequently 'dropped'.

It is clear from the Birmingham case that the use of the SPZ was at variance with central government's intentions and that far from being 'simplified' it was in fact more regulated than the erstwhile regime. The following will address the two main research questions set out in the research methodology: the use of the zone and its influence.

The use of the zone

Saltley was a semi-derelict industrial area with large vacant sites and contaminated land – the ideal site for a zone according to PPG 5. Although physically suitable it is questionable whether the area was appropriate beyond the narrow physical confines of the PPG. The juxtaposition of the zone to high-density residential areas highlighted the confusion of purpose for the SPZ between the City Council and the DoE. The latter pushed for more uses to be permitted by the zone. However, the more uses are included in a zone, the more uncertainty is created. This inevitably leads to more conditions being attached and a greater overall complexity. Council officers are in no doubt that if the zone had restricted itself to the uses originally proposed (B1, B2 and B8) then the zone and the area itself would have been far more suitable.

Prior to the zone there was no development plan in the area and there is evidence to suggest that the City Council saw the SPZ as an opportunity for a loco or surrogate development plan. However, there was no suggestion from the DoE or anyone else that there were problems of uncertainty because of the lack of a development plan. In a similar vein there is no evidence of an inflexible attitude towards development in the area outside what the draft zone would have permitted. All the uses permitted in the draft zone (apart from A2) were present in the zone anyway and although the zone limited B2 uses adjacent to residential areas the draft document pointed out that B2 uses may well be acceptable in such locations but should be subject to normal planning procedures.

Given the situation it is unclear why the DoE wanted to make the scheme more flexible in its range of uses. In effect the DoE were attempting to impose a regime that went beyond what the market in the area had determined as suitable (the City Council had made it clear they would welcome virtually any uses in the area).

For both the DoE and the City Council the Saltley zone was a means to an end – although these ends differed. From all the evidence it is not clear what the DoE thought the zone would achieve and why the area was suitable for an SPZ. Birmingham City Council on the other hand had taken on the SPZ as a condition of the Secretary of State withholding a UDC for the area. According to the Council they would not have pursued it in any other circumstances. Nevertheless, the Council were not vehemently opposed to the zone and accepted that if they were to adopt it they might as well make some use of it. Their aims differed from PPG 5 and included guiding investment and filling the gap left by the lack of a development plan. Beyond this the Council consented to 'going through the motions'. This is perhaps best demonstrated by the decision to drop it as soon as it was no longer needed. In all the Council preferred its own Delegated Authority Zone approach where powers were given to officers in conjunction with the Chairman of the Planning Committee to determine applications in certain areas. This led to quicker decisions and was used by the Council in a number of run-down areas.

The influence of the zone

The confusion over the purpose of the zone led to a scheme that was practically unworkable. The consensus of opinion within the Council and the Development Corporation was that the scheme, by allowing up to five uses, actually created uncertainty in the area. This uncertainty was ameliorated by the number of conditions attached which went far beyond what would have been attached to a normal permission. Beyond the practical usefulness of the zone there were other benefits. It focused attention on the twelve types of grants available. It also provided a development control framework for the area even after it was decided not to proceed with it.

The Slough zone

Slough is a town of around 100,000 people located to the west of London, close to the M25 and just off the M4. The town grew rapidly during the 1970s on the back of commercial activity associated with Heathrow airport. Berkshire as a whole has built up a reputation for being a prime location for high-technology firms, though Slough itself has not benefited as much as other nearby towns in this respect. The town has consistently suffered from above-average unemployment and light industrial employment (MacGregor *et al.* 1987). In order to try and combat this the town's district-wide plan (adopted in 1992) argues for a relaxation of structure-plan limits on growth in Berkshire.

Slough trading estate is located close to the town centre and was the first and largest trading estate in the country. Within its boundaries there is some 7.5 million square feet of business premises occupied by around 450 companies. The owners of the estate, Slough Estates, approached Slough Borough Council in early 1989 about the possibility of adopting an SPZ in order to give them greater flexibility in meeting market demands. They had been closely involved in formulation of the SPZ policy and their Chairman, Sir Nigel Mobbs, was keen to demonstrate his commitment to the idea. Slough Estates wanted to be able to adapt quickly to market demands for different uses of their buildings and Slough Borough Council (SBC) were happy to support them especially in relation to their wider desire to encourage greater employment opportunities for the town. It was agreed that Slough Estates would prepare the scheme in close collaboration with the Council.

The scheme included a blanket B1, B2 and B8 approval across the estate with some limitations in sensitive sub-zones adjacent to residential areas. However, SBC and the DoE (who had been consulted early on in the process) recognised the potential for traffic problems with a large amount of B1 office use in the area. Slough Estates proposed to overcome this by sub-dividing the B1 use class into B1(a) (offices) and B1(b) (light industry), the scheme only allowing the latter. However, the DoE objected to this as it effectively made distinctions within a single use class which the new 1987 Use Class Order had sought to eradicate. A possible way around this, the DoE suggested, was a blanket B1 approval combined with an upper limit of 20 per cent of the estate.

Regardless of this problem, all parties agreed that a traffic impact study was necessary and the close involvement of the County Council was crucial. Berkshire County Council had already made their views known on the scheme and objected not only to the highway implications but also the potential over-heating of the local economy that the structure plan aimed to avoid. If these matters could not be resolved then the county made it clear that they would force a Local Inquiry into the zone that would delay it substantially. In addition to these problems the Borough Council's Environmental Health Department had insisted on a number of conditions to the zone including noise limits which Slough Estates considered unreasonable. Slough Estates commissioned an independent assessment of the highway implications of the zone to gauge the likely contributions required by the county for highway improvements. The study concluded that these would be quite high (around £5m) and outweigh any benefits the company might derive from the zone.

The problem of a potential county objection was tackled by the Chairman of Slough Estates, Nigel Mobbs, who met the Prime Minister on 4 June 1990 to discuss the problem. This appeared not to help. Given the county's stance the alternative was to withdraw the B1(a) aspect of the zone altogether. This decision corresponded with a downturn in the property market and reduced pressure on Slough Estates to proceed as quickly as possible with the zone. The zone was eventually adopted in 1995.

A number of important factors emerge from the Slough Zone which are relevant to this study. First, the use of the (then) adoption procedures by the county to exert their interests demonstrates the lack of freedom which appears to go against the government's aims for SPZs. Second, the concerns of the erstwhile planning regime were clearly influential on the zone as it progressed through its adoption procedures and questions the role of SPZs as a 'planning-free zone'. Third, the need for the zone to conform to the existing development plans answers one of the specific criticisms of the SPZs, that they are a 'black hole' in the development plan system. Fourth, the role of Nigel Mobbs in attempting to influence the policy formulation process especially after the delays caused by the County Council tells us something about the influence of the private sector upon the changes to planning during this period.

The use of the zone

The Slough Estates zone – an industrial estate in need of renewal – met the physical suitability criteria of PPG 5 though there are question marks over the suitability of the area beyond the immediate concerns of PPG 5. The zone as proposed would, in the eyes of the County Council, have been contrary to the Structure Plan in terms of encouraging further office growth and traffic. In pursuit of this objection BCC used the adoption procedures for the zone to remove the B1 aspect. The adoption procedures have since been modified to remove the inevitability of an inquiry following objections. Beyond the concerns of the erstwhile regime there were also question marks over the surrounding area's suitability. Residents expressed concern over the extension of B2 uses and this was supported by the Council's Environmental Health Department. This situation led to complicated sub-zones and conditions being attached that limited the extent of B2 uses. Interestingly, B2 uses had been permitted close to residential areas prior to this though they had been restricted by suitable conditions. It was the flexibility of B2 uses that could occupy the area (the same flexibility that the government sought for SPZs) that led to uncertainty for adjoining residents. The conditions attached to the zone were considered excessive by Slough Estates and far beyond what would be attached to a normal permission.

The preconditions of a zone as set out in PPG 5 – inflexibility, uncertainty and delay – were no worse in Slough than in any other area, and in some cases were significantly better. However, the Structure Plan did impose some restrictions in the form of constraint policies though the Local Plan gave the estate more flexibility in this respect. In addition to this there was also a degree of political uncertainty especially following the 1990 elections which returned a larger Labour majority. It was the support of the new Labour members that the local residents sought in order to push their case for greater environmental protection.

The aims of the zone appear to go beyond the physical advantages. The role of the Chairman in the formulation of SPZ policy was clearly linked to the decision to pursue a zone both in Slough and at the company's other large industrial estate in Birmingham. Of all the zones being pursued in the UK only two have been initiated by the private sector and both of them involve Slough Estates. The removal of the B1(a) element of the zone severely limited its effectiveness, though having spent £300,000 on preparing the zone the company decided to pursue it to adoption. If the company had known from the start that B1(a) would not be included then they would not have proceeded with it.

The influence of the zone

Although only recently adopted the zone has had influences, though those were mainly prior to adoption. Probably the main influence to date has been the use of the zone as a vehicle for different and sometimes conflicting interests to pursue their own ends. In this role the zone has acted exactly like the erstwhile regime it was meant to simplify. The zone also raised the question of environmental concerns from nearby residents. While not being moved to object to single applications the prospect of a blanket permission raised awareness of the situation. According to Slough Borough Council any future applications will now have to conform with the higher standards imposed by the zone. The issue of highways was also raised and now the County Council will be expecting large-scale contributions rather than one-off improvements related to a single application.

The Derby zone

As the first SPZ in the country the Sir Francis Ley Industrial Park in Derby has attracted a good deal of attention from practitioners, the professional press and academic studies (Lloyd 1987; Cameron-Blackhall 1993). In the nineteenth century Derby became one of the principal industrial centres in the East Midlands and grew to a population of around 210,000. In recent years unemployment in the city has risen as many of the older industrial employers have closed. The Leys site is located to the south of the city within Babington and Litchworth wards – areas that have consistently suffered from higher levels of unemployment and poorer quality housing than the rest of the city.

Throughout most of the 1980s the Labour-controlled Derby City Council (DCC) followed an aggressive pro-economic development stance which was quite successful in attracting a number of in-coming firms including Toyota. Following elections in 1988 the Council became hung and attention switched to more environmentally based works including improvements to housing stock.

The Derby zone was an initiative of the former Director of Development, Ian Turner. A property company, J. F. Miller had bought a derelict former foundry and approached the Council for financial assistance to help with this in late 1987. The Council, who were very concerned about the economic decline of the city, saw an opportunity to achieve some much-needed regeneration. The government had recently announced the amalgamation of the Urban Development Grant, Urban Regeneration Grant and Derelict Land Grant into the City Grant. To qualify for this grant it would be necessary for a developer to prove that a project would not proceed without it and would involve investment of £200,000 or more. Authorities would, in effect, be in competition with each other for the funds.

Both Millers and the City Council realised that the land clearance including removal of the contaminated material would require City Grant funding to make it viable. To get an edge on other authorities Ian Turner suggested combining the grant application with a commitment to an SPZ – recently announced by the government. In addition, being the first zone in the country would bring a lot of publicity and interest not only in the site but also to the Council. The common view and agreement on the need to move forward together meant that Millers abandoned some of their original ideas for the area including possible retail uses. The DoE were very keen and encouraged Derby to proceed apace. Informally, the DoE told Turner that an SPZ would virtually guarantee the City Grant.

The draft scheme was prepared based on B1, B2 and B8, *sui generis* and some open storage uses. The scheme was agreed and placed on deposit on 4 May 1988. As DCC had already carried out extensive consultations with statutory bodies prior to this it did not expect any objections. On the last day of deposit a petition was received from 11 local residents. The residents were concerned with the secondary/emergency access onto Coronation Street which, they believed, was already over-congested. If the petition were not withdrawn a public inquiry would be necessary which would jeopardise the City Grant. DCC responded to this in two ways. First, it sent officers to each objector's home to persuade the residents that it would be in their interest to allow the site to be redeveloped rather than stand derelict. Millers agreed to amend the scheme and limit the access onto Coronation Street.

Second, when all but one resident withdrew their objection and this resident asked for planning permission for a garage in return for withdrawing his objection Ian Turner wrote to him personally. Pointing out that planning permission can only be granted if formally applied for, he nevertheless agreed the garage was acceptable in principle. The resident withdrew his objection.

As the first SPZ in the country the Leys zone received a lot of publicity with a DoE minister visiting the site and announcing a £3.28m City Grant the day after the scheme was adopted on 3 August 1988.

Behind the headlines of Derby's achievement in being the first SPZ in the country is a picture of an authority prepared to use a mechanism they would not normally consider to achieve their aims. Clearly the zone had far wider implications than its land-use impact.

The use of the zone

According to PPG 5 the Leys zone was exactly what the government had in mind for an SPZ. Its derelict industrial nature typified the problems being experienced by many areas during the 1980s (Derby City Council 1985) and the Council's desire to redevelop it and promote business and jobs echoed the sentiments if not the rhetoric of government advice. However, all of the parties involved agreed that the scheme would not have proceeded without the City Grant link. It is probably coincidental that the City Grant that accompanied the zone is the largest to date though Ian Turner believes that the SPZ was made a prerequisite of the grant. The link is not quite that simple though. A number of officers at the Council pointed to the inverted political 'nose-rubbing' of a Labour authority implementing a Conservative government scheme. In terms of the physical relationship of the site to the adjoining residential areas, many of the residents' fears have come true. Congestion has increased on local roads and the Council admits that had the proposal been progressed through the normal planning procedures this could have been tackled by a legal agreement securing off-site highway works.

None of the preconditions set out by the government – inflexibility, uncertainty and delay – could be argued to have been present in the area prior to the zone. There is little or no evidence that the planning system in general or the particular regime of Derby was regarded as being inflexible enough to warrant an SPZ. The Local Plan zoned the area for industrial use and Millers had no concerns about the inflexible attitude to the redevelopment of the area. Indeed Millers believed that the inflexibility of uses was far less important than the flexibility to alter building size.

The two main players in the zone, DCC and Millers, had similar aims for the Leys SPZ. Both saw it as a means to an end though in a different way from the government. While the government saw SPZs as a mechanism to facilitate development it was used as a means of securing City Grant funding to facilitate development. Nevertheless, both Millers and DCC realised that if they were to adopt a zone it may as well be useful – besides, the DoE would not have accepted a zone that was simply 'going through the motions'. In this respect the usefulness of the zone has been limited. DCC point to the promotional value while Millers believe there was some flexibility in unit design.

Influence of the zone

Apart from the City Grant issue the actual influence of the zone is doubtful. Officers at the City Council believe that a similar application on the site would have been granted permission in a much shorter time and the minor modifications procedure could have been used to alter the design and layout without resorting to further applications.

Overall, neither DCC nor Millers places too much emphasis on the role of the zone in the success of the scheme. A general lack of similar units in the area available for renting or buying meant that the Leys estate would probably have been commercially successful anyway. The fears of some of a 'planning free-for-all' were limited by the restricted range of uses, conditions, the existence of other legislation (e.g. building and environmental health control) and restrictive covenants. Nevertheless, in the rush to adopt a zone and secure City Grant funding the obvious losers were local residents.

The Cleethorpes zone

Cleethorpes, like its larger neighbour Grimsby, grew up as a fishing town on the Humber Estuary. The decline in this industry since the early 1970s has been partly offset in employment terms by the growth in food processing, construction and transport-related jobs. Nevertheless, the town has been designated a Development Area where assistance is available for projects which provide or safeguard employment. In addition to the deterioration of fishing the increase in foreign holidays has led to a gradual decline of the town as a holiday destination, especially the traditional centre of this industry, the North Promenade. This area of the town extends for about half a mile from the town centre west along the coastline and is dominated by amusement arcades, gift shops, night clubs and public houses. Generally, the area is recognised as being physically decayed (Cleethorpes Borough Council (CBC) 1987, 1988a) and in need of attention. British Rail, a major landowner through its station and marshalling yards, has for some time been looking for an opportunity to pull out of the area and close the station.

The Borough has a fractious political history of alliances and splits within ruling groups. Throughout the 1980s a Labour/Official Conservative alliance ruled the Borough. The Official Conservative group had split from the Conservatives over a disagreement concerning national policy – the former following a more local 'common sense' approach. The resulting alliance was characterised by weak political leadership and a policy vacuum for officers. In terms of land-use policy the coalition could not agree on a direction for the Local Plan. The Official Conservative group felt that restrictions on parking and access were necessary to improve the attractiveness of the resort, leading to a disagreement with the Labour group. As a consequence the 1965 Cleethorpes, Humberston and Waltham Town Map drawn up in the heyday of the resort was the statutory development plan for the North Promenade and zoned it for railway purposes, amusement, recreation and entertainment. Although other plans have been prepared for the area, none had been adopted because of political disagreement concerning aims.

Officers at the Council realised that the North Promenade area was in particular need of redevelopment and the lack of an up-to-date statutory plan was a real problem (CBC 1988). As numerous attempts had been made through 'normal' Local Plan channels senior officers at the Council agreed in November 1987 that a different approach would be needed. One possibility raised was an SPZ. This, they believed, could target the North Promenade and by combining plan and permission provide the necessary certainty that the political situation lacked.

The idea was presented to members in February 1988 and they agreed to proceed with it. Simplified Planning Zones were presented to the Planning Committee as a tool to speed up regeneration while maintaining development standards.

Different Sections within the Council had already come into conflict over their different interpretations of the zone. The Economic Development Section was pushing for a wide range of uses with as few restrictions as possible while the Planning and Environmental Health Sections realised that controls and restrictions needed to be maintained especially as the Promenade was bounded by residential areas. These two very different views of the zone could not be easily reconciled and the draft scheme steered a path between regulation and deregulation.

The draft scheme contained a wide range of uses including A1, A2, A3, C1, C3 and D2 accompanied by 14 conditions and a further set of restrictions covering parking and noise. This approach provided for deregulated uses which were regulated by a high number of conditions. Six additional conditions were added following public consultation on the draft scheme. Although the SPZ was generally welcomed by those consulted the County Council expressed a number of reservations. First, they questioned the assumption that the zone would be developed by one developer and pointed out the problems that would emerge if it proceeded piecemeal. Second, the county were concerned that the proposed retail uses in the zone were contrary to the Structure Plan which sought to concentrate such uses in the town centre. The Borough Council made a number of alterations to the scheme as a result of this including limiting retail and residential uses, the materials to be used in construction and putting a limit upon the type and amount of substances that could be stored in the area.

Pivotal to the scheme was the stance of British Rail (BR) who were required to relocate the station to allow development to proceed. On this assumption Bellway Urban Renewal had put together a proposal in association with the Council and in accordance with the provisions of the draft scheme. The ability of BR to pay for the relocation was closely tied up with money derived from redevelopment (as they owned most of the land). However, the viability of the scheme began to be questionable for BR with the on-set of the property recession. In addition, the 1990 Planning Act required Local Planning Authorities to adopt district-wide plans for their areas. This would force the Council to agree on objectives for such a plan and undermined the need for an SPZ. Both factors conspired to lead to the downfall of the zone.

The Cleethorpes zones appeared to follow the government's aims for SPZs though it is not clear from the above that the uncertainty that justified it did not derive from the planning system. What is also evident from the Cleethorpes case is the confusion that arose out of the lack of clear local objectives.

The use of the zone

The North Promenade met the physical suitability criteria of PPG 5 – it was a mixed-use area in decline. However, the close proximity of the zone to residential areas limited the practical scope of the zone because of the need to avoid any conflict. Beyond PPG 5 there were clearly other physical restrictions upon the site including limited access and the need to relocate the station to facilitate development. It was these problems which, the Council believed, led to the need to impose a high number of conditions which limited the usefulness of the zone. This demonstrated the need for any zone to relate to its surroundings and not be a 'black hole' in the development plan system.

Obviously, uncertainty was a precondition for the zone, though this related more to the political environment than to that of the planning system. Ironically, as the Council pointed out, the lack of an up-to-date Development Plan for the area meant that local firms believed they would receive planning permission for most things. However, there was uncertainty for investors in the area who preferred the certainty of a land-use framework – this was not helped by the SPZ. Although the SPZ provided a framework for decision-making potential investors preferred the erstwhile system that gave them a 'piece of paper'.

The aims for the zone especially related to those in PPG 5. For political reasons uncertainty was omitted although this is recognised as being the most important justification. As the purpose of the zone was left deliberately vague to avoid political problems this led to a lack of objectives to guide the scheme and led to the problems of different interpretations and aims described above. The other problem, as the County Council identified, was the dependence on one developer who had all but dictated the scheme. If this deal fell through then it would have been difficult to coordinate piecemeal development.

The important difference between the certainty envisaged as an aim for SPZs by the government and that by the Local Authority was its origin. The government's uncertainty lay in the planning system *per se* while the Cleethorpes uncertainty derived from the political environment.

The influence of the zone

Although not implemented the draft zone had a number of effects. First, it gelled political, officer and local support for the future of the North Promenade. Although uncertainty over the land-use framework existed throughout the Borough, the North Promenade was used for a zone as it suffered the most acute decay, was a relatively small area and did not excite a great deal of political controversy. Second, the zone gave direction to the area. Political uncertainty had led to incremental decision-making in the North Promenade as well as other areas. The zone, even its draft form, specified the uses that were considered acceptable, allowing companies to plan within a land-use framework. Third, the zone was a basis for negotiation with BR over their relocation – and without it BR would not have been even willing to discuss the matter. Fourth, the zone furthered the ends of the County and Borough's transport strategy by seeking to relocate the station. Finally, the zone generated publicity for the area and, in particular, advertised the potential for redevelopment opportunities that existed.

Conclusions

The four case studies have questioned the assumption that Thatcherite policy was translated into Thatcherite policy outcomes. The case studies also show that in the field of SPZs there was considerable scope for autonomous local action which derived from a lack of directed policy objectives and the discretion offered by the zone legislation. This allowed local authorities to substitute their own objectives which were at odds with the spirit of SPZs and the central tenets of Thatcherism. Beyond these general comments I will also address:

1 the influence of the zones, and
2 offer some explanations as to why and how SPZs have been used for reasons at variance with their Thatcherite aims.

Each of these will now be examined in turn.

The influence of Simplified Planning Zones

The influence of SPZs has two aspects: first, in relation to the government's aims and second, to those at the local level; and these, as we have seen, differ. Overall there have been very few attempts to adopt an SPZ and even fewer have actually been adopted. The Ove Arup study (DoE 1991b) put this down to a number of reasons including the lengthy adoption procedures and the property recession. This is itself a testament to the influence of SPZs given that they could feasibly have been adopted in the vast majority of local authorities across the country. Nevertheless, in those zones that have been adopted development has taken place, though as we have seen this disguises the true influence of the zones. In terms of the government's aims three factors are central to any SPZ: physical suitability, preconditions and aims.

Physical suitability

1 All of the zones met the physical suitability criteria of PPG 5, which were deliberately drawn widely to present SPZs as an extension of the government's wider policy of deregulation. But the wide scope of the zones' suitability in the government's eyes reflected a lack of objectives and had several consequences in the case studies.
2 There was considerable scope for others, particularly local authorities, to substitute their own objectives and use an SPZ as a means to achieve them.
3 The lack of central objectives led to problems with adjoining uses.
4 The need to impose high numbers of conditions and complex zoning arrangements on 'physically suitable' though inappropriate areas conflicted with the spirit of deregulation within the zones.
5 Confusion over the government's objectives led to attempts at general deregulation through the number of uses permitted rather than more specific schemes tailored to the needs of the area.

The physical suitability aims of the government lacked wider objectives. It therefore fell to those setting up the zones to use them as they thought fit. Without such objectives the zones had to depend on local direction (which in the four case studies was at variance with the government's wider deregulatory philosophy). All the studied zones, in substituting their own objectives, went beyond the regulation found in the erstwhile regime.

Preconditions

The government believed that SPZs would overcome uncertainty, inflexibility and delay in the development process. A number of MPs and consultation responses had raised the ambiguity of these terms and had sought to link them with objectives for the zones. The studies showed that different types of uncertainty existed in different zones not all of which were land-use-related and not all hindered development in the area.

In Cleethorpes the lack of a development plan led to certainty in the eyes of local firms of what would be allowed – it was political uncertainty that had led to the zone. Similarly, Birmingham had no development plan but the survey suggests that this was not perceived as creating an uncertain environment for developers. In some ways zones actually created uncertainty. A number of those involved pointed to purchasers wanting a 'piece of paper' to prove permission had been granted in order to sell land. Developers also still went to the local authority to check whether permission was required for individual applications; some investors were unsure about the legality of the scheme and in the Slough case the zone was actually contrary to the Structure Plan.

The flexibility the government thought the zones would offer was also elusive. In all the zones the uses permitted by the scheme would probably have been permitted under the erstwhile discretionary regime. However, as was clearly demonstrated in the Birmingham zone, the general aim of 'flexibility', i.e. a wide range of uses, as pursued by the DoE regional office led to the need to impose a wide range of conditions and sub-zones. Deregulation as an aim in itself (an attempt at 'non-plan') was insufficient, as the City Council pointed out. What was required was a scheme tailored towards the specific needs of the area; otherwise a wide range of uses created uncertainty. Again, it was not clear from the government what the zones sought to achieve, which in turn left question marks over what sort of uncertainty or inflexibility the government had in mind. It was therefore possible for the DoE and Birmingham City Council to argue that their schemes, one based on deregulation and the other on uses already established and permitted in the area, were both consistent with PPG 5.

Aims

All the study areas sought to promote development of some sort though did not necessarily use the SPZ to achieve this as the government envisaged, i.e. through deregulating the planning system. Birmingham wanted to promote redevelopment through its Urban Development Agency while maintaining control and used the zone to successfully fend off an Urban Development Corporation. Cleethorpes needed a land-use plan to act as a focus for developers. Political disagreements meant that they could not get a Local Plan adopted and so they used an SPZ instead. Controls within the zone were maintained by conditions and sub-zones. Derby needed an SPZ to secure their City Grant though its land-use provisions were of little practical use. Slough Estates' hope to use an SPZ to avoid highway contributions had less to do with deregulation and more to do with financial gain.

Birmingham, Derby and Cleethorpes would not have used an SPZ unless it was a necessary way of achieving wider aims and all agreed that if the zones had been adopted the benefits in terms of the government's aims would have been minimal. In all zones the number of conditions and sub-zones meant an overall increase in regulation rather than the government's intended deregulation.

Although the influence of the zones in terms of national objectives was limited, all the study zones (because they had been used for reasons other than envisaged by government) had significant local influences related to their local objectives. The various monitoring reports on the Derby zone questioned its ability to promote redevelopment beyond its marketing potential. By far the most important aspect in the site's success, according to the City Council, was the City Grant and the lack of availability of alternative sites nearby. The zones did not speed up the development process and the flexibility offered was no more than would have been permitted anyway. Beyond these influences the zones have been successful in fending off an Urban Development Corporation in Birmingham and providing a loco development plan for Cleethorpes North Promenade – neither of which was seen by government as the aims of SPZs.

Why and how SPZs have been used for reasons at variance with their Thatcherite aims

If SPZs did not achieve their Thatcherite aims and, as the studies show, proved an increase rather than decrease in regulation, the question that then arises is: how did this come about? As the introductory chapter sets out, the government approach to implementation gives an indication that policy failure is due to a self-inflicted implementation gap. Implementation problems are not unique to Conservative governments though they were uniquely severe because they insisted on an inappropriate (and ill-conceived) model of implementation. Looking at Sabatier's (1986) list of preconditions for effective implementation we can see how the government failed in its approach to 'top-down' implementation.

Clear and consistent objectives

One of the points that arises throughout this work is the lack of any wider objectives for SPZs beyond promoting development through deregulation. This lack of clear and consistent objectives left SPZs in a policy vacuum and allowed others to fill the vacuum with their own uses as we have seen.

Adequate causal theory about the situation and its causes

Many writers believe that EZs and SPZs were ideological in their origins and developed from a general Thatcherite philosophy that applied blanket prescriptions to very different situations. In terms of SPZs there emerge two related aspects. First, the government believed that deregulation would create certainty for developers as they would no longer need to apply for planning permission. Any certainty offered by the scheme created uncertainty for those adjoining the scheme. This leads us to the second aspect – that to create certainty through deregulation the zones needed to include a large number of conditions beyond those that would have been imposed under the erstwhile discretionary regime.

Appropriate policy tools and resources

The inadequate theory as described above led to an inappropriate tool to encourage development. Local authorities and the firms in their areas recognised that factors such as interest rates, the recession and site costs were more important than planning regulations. SPZs and deregulation provided no extra benefits to their efforts, which is one reason why they were so unpopular.

Control over implementing officials

As the main implementing agency (the Secretary of State has not directed any zones thus far) local authorities could use the considerable discretion offered by the zone legislation and the lack of any clear central objectives to use the zones as they and their politicians saw fit. In these circumstances where conflicts between central and local government did arise, as in Birmingham, the lack of central objectives meant that two different views of the purpose of SPZs could be held.

Support of interest groups and agencies affected by the policy

The conviction politics approach of the government generally sought to exclude many interests typically included in the formulation of policy. In terms of SPZs it is clear that the government formulated policy in association with Slough Estates, a large private landlord, and excluded organisations such as the Association of District Councils (ADC), Association of County Councils (ACC), Welsh Office and others until formally consulted in 1984. These interests, although by-passed by the Thatcher government, have continued in existence and acted as a major constraint on the development and implementation of radical policy.

Stable socio-economic contexts which do not undermine political support

The most important socio-economic change that affected SPZs and the government's wider policy of property-led urban regeneration was the recession at the end of the 1980s. We have already seen that there was little support for the concept regardless of this. Nevertheless, others have demonstrated the fragility of this property-led approach during recessions and there is little reason to suspect that these effects would be any different for SPZs (Healey *et al.* 1992).

In addition to these 'top-down' implementation problems SPZs also demonstrate the role of some 'bottom-up' implementation failures: in particular, the role of 'street-level bureaucrats' in using their own objectives, e.g. the difference in approach between different officials in Cleethorpes. Evidence pointed to the existence of and conflict between policy sectors within and between organisations and the view that actors within a bureaucracy can pursue disparate objectives within an institutional framework is backed up particularly by the Derby study.

The implementation perspective offers a powerful explanation of why many of the Thatcher government's objectives were not achieved as envisaged. This 'self-imposed' implementation gap offered the opportunity for autonomous local action and politics to use the SPZs for their own ends.

7

THE NEW RIGHT'S COMMODIFICATION OF PLANNING CONTROL

Mark Tewdwr-Jones and Neil Harris

Introduction

The notion of New Right ideology within the planning system found its expression in the 1980s under a strong government. The New Right ideology, as applied by the Major administrations in the 1990s found expression through far more subtle manifestations of changing procedures and initiatives. Thornley (1993) identifies Circular 22/80 'Development Control – Policy and Practice' as the centrepiece of development control policy of the first Thatcher administration. However, no such 'centrepiece' is so readily agreed upon in the context of the 1990s. This may be considered to reflect a series of potential issues. After 1990, the government may have placed more reliance on incremental alterations to development control policy and procedure. Alternatively, we might hypothesise that development control as a policy process had been comprehensively 'dealt with' in the 1980s and now represents the baseline of the Thatcherite planning model. Or, does the lack of a centrepiece development control policy merely reflect the advancement of New Right ideology into a new phase, by a focus on procedures rather than policies? Thornley (1993: 222) identifies the importance of procedures within the planning system but suggests that this was actually a 'second dimension' to the Thatcherite reforms. The changing political and economic conditions of the 1990s could have forced the government to now concentrate on this second dimension of planning.

We commence this review of development control as it has been affected by New Right ideology in the 1990s by considering a brief three-way analysis of the process. An analysis of these categorising components, legislation and policy initiatives, will lead us to contend that the greater emphasis in the 1990s has been on initiatives rather than policy as in the previous decade. By considering the changes in this context, we wish to argue that the period since 1990 has seen a change in approach of New Right ideology within planning, towards instrumental means.

We wish to discuss three principal arguments relating to the role and operation of development control in the UK in the 1990s. First, we consider that the New Right focus towards development control by the Major administration was on improving the procedural efficiency of the system and to reconstitute development control as a public-service ethic within local government. The most noticeable aspects of this change were in the requirements of both the Planning and Compensation Act 1991 (in development control decision-making) and the Citizen's Charter (in fostering quality and efficient procedures).

Secondly, we consider that the government became more interested in setting market-oriented performance-criteria standards for the operation of development control and thereby standardised planning control across local spatial scales. This has inevitably led local planning authorities to become concerned more with the process of development control through administration and commodification rather than with utilising the process to achieve the most desirable and highest-quality planning outcomes.

Finally, in parallel with the pursuit of efficiency nationally, we have identified a backlash in the operation of development control within certain local areas against the imposition of a standardised planning system that is not perceived as meeting local socio-economic needs, but which central government has labelled 'malpractice'. Although there are a number of other matters relating to development control in the 1990s that we could have discussed, we believe these three arguments are the most pertinent, and are far the most revealing in theorising the current and future transformation of the planning control process. These three strands are addressed in this chapter.

Defining issues and concepts

The concept of 'the client' in development control is one that best identifies the changes that have occurred to planning between the 1980s and the 1990s. An analysis of the clients of development control can be related just as much to issues of administrative efficiency within local government generally. It is not the purpose of this chapter to look at the changing nature of local government or at the changing relationships between central and local government; however, an examination of planning in the 1990s will by its very nature necessitate consideration of matters that extend beyond the planning function. In particular, the associated issues of performance indicators, client-orientation and the public-service ethic are all evident within planning control in the UK in the 1990s. These components are evident in many public-service areas; they are not confined to planning nor have they been developed exclusively as a central plank of 'planning policy ideology' of central government.

Issues of quantitative efficiency within public administration have been well considered within the literature, possibly because of the interests of the work of both the Audit Commission and the government (through its Citizen's Charter) to legitimise a quantitative assessment of 'performance' by agencies of governance and public policy researchers. In fact, we would go so far as to suggest that quantity has been deliberately reformulated to be almost synonymous with quality. This line has developed further, and with some degree of importance within planning in the 1990s, because of the procedural orientation of planning changes since the introduction of the Planning and Compensation Act 1991. Research that considers the issue of quality within public services, however, is less apparent. This could be because quantity and quality issues cannot very easily be dealt with through the same mechanisms; after all, attempting to commodify efficiency within planning can only partly be achieved through a 'remote-control mechanism' of performance criteria assessment. Within the scope of this chapter on development control and the New Right, therefore, we have chosen to consider quantity and quality components within the planning system of the 1990s, although we are extremely aware that it is by no means a straightforward task.

Collecting data to identify performance and good practice and subjecting that data to statistical analysis is a relatively straightforward method of assessing a local planning authority's abilities at planning control. However, as statistics they may not reveal much in themselves; they may also mask a great deal of circumstantial evidence behind performance; they may also portray a local planning authority in an unfair light. There are therefore problems with quantitative assessment. Qualitative assessment of planning control functions of local authorities provide far more interesting information of the efficiency of planning practice, but identifying a local planning authority that could be thought of as 'representative' of the general process across the country may not be easy. In fact, qualitative assessment of development control within named local planning authorities in the 1990s has been associated with reports of malpractice, irregularities and misrepresentation (Department of the Environment 1993a; House of Commons Welsh Affairs Committee 1993), although, uniquely, these studies reveal far more about how the development control service is operated today than collecting and assessing aggregate statistics. Within the context of the theme of this book, in identifying local action against the imposition of a national standardised process founded on New Right principles, qualitative elements rather than quantitative criteria have provided researchers with a more readily identifiable assertion of different and locally politically divergent practice within development control.

These well-publicised local planning authorities have been subjected to a great deal of criticism for their alleged malpractices within development control. However, no evidence of corruption has been found by the official inquiry established to look at the workings of each authority; we might even go on to suggest that the 'malpractice' label has been tagged by central government onto those local planning authorities that are not operating development control along the lines established at the national all-Britain level. These local authorities have rather shaped and reshaped the planning control process in their localities to meet their own socio-economic and political needs, and the fact that this remains possible after the New Right reorientation of planning in the 1980s is indicative of the resistance of planning control to conforming to (strong) top-down control.

There is a further matter that all these publicised cases have in common: they are wholly or largely rural authorities. Thornley's (1993) assessment of the impacts of Thatcherism on planning practice invariably focused on an explicit urban context. Thatcherism is a market-led ideology, and such ideologies by their very nature are guided predominantly by urban issues and affairs. The development of government policy within the Thatcher administrations that affected the planning system was undoubtedly urban-guided, but this was also the same national standardised framework within which rural planning occurred. Academic assessment has easily identified Thatcherite impacts upon planning as essentially urban, but there has been an associated rural context. The most extreme forms of New Right planning were implemented in the 1980s through such urban measures as Enterprise Zones, Urban Development Corporations and Simplified Planning Zones, where development control powers were either removed or placed in the hands of agencies of governance other than the democratically accountable local authorities. However, to focus on these urban initiatives alone is to mask the changes that have also been occurring in rural areas, albeit in a different form.

The Thatcher governments protected the best of the landscape and vigorously upheld countryside designations. This was the principal effect on rural planning of New Right ideology; it was not so much an 'effect', but rather a 'non-effect' – radical planning changes in the countryside along Thatcherite lines was simply not an option for the Conservative Party, many of whose votes were shire residents. But we should not fool ourselves into thinking that there have been no knock-on effects of the ways in which the Thatcher and Major governments changed the policies and procedures of planning in rural areas. If anything, and if our contention that the focus of the post-1990 ideology towards planning was in procedures rather than policies is correct, it is only since 1990 that the possibility of rural discontent with the operation of a Thatcherite planning process would surface. There has been no especial urban bias of planning in the 1990s; the changes that have occurred have affected rural areas as much as the metropolitan areas. We shall go even further by suggesting that the 1980s saw the neglect of rural planning issues by the Thatcher governments, and the frustration within some rural areas towards this neglect actually fostered the possibility of local discretion occurring against the imposition of a 'hands-off' control focused on urban initiatives and markets.

These are contentious issues of course. And we can only begin to theorise such matters within the confines of this chapter. But they are worthy of further consideration and we shall address these arguments as they have affected and have been affected by development control in the 1980s and 1990s. Our principal arguments are that:

1 The New Right ideological focus of the post-1990 Conservative government towards planning control was centred upon changing procedures rather than policies.
2 Central government's market-oriented pursuit of setting performance criteria checks in development control has affected both urban and rural areas but has added nothing to ensuring quality outcome.
3 There has been an apparent backlash in certain rural areas against the imposition of a planning system founded upon New Right ideology and this has been most readily identifiable within qualitative assessments of local development control operations.

These arguments are structured into the following sections of the chapter. The initial section outlines the nature of development control in the 1990s, by examining the legislative changes, the policy changes, and the introduction of other control-centred initiatives. This will enable a thorough assessment of the 1980s–1990s differences, and to outline how the government has turned its attention to improving the efficiency of the procedures of development control. Following this argument, we examine the most notable changes to the development control system by examining, first, the increased importance of costs and performance criteria and, secondly, the introduction of an enhanced public-service ethic through the Citizen's Charter. The second half of the chapter looks at the operation of locally divergent development-control practices within selected local planning authorities in rural areas of Britain, and an attempt is made to theorise the reasons for these practices and to suggest how a New Right ideology may have contributed to these 'malpractices'. We conclude the chapter by returning to our three hypotheses and by suggesting what the future directions of the development-control service might be.

Development control in the 1990s

Thornley (1993) provides a detailed and thorough account of development control changes during the Thatcher administrations through an analysis of the various legislation and other instruments issued and enacted by successive governments. Our contention is that changes to the development control system post-Thatcher have been incremental and opaque in comparison to the Thatcher years. It is evident from the following assessment that various important strands of Thatcherite planning continue to be developed and perpetuated through recent instrumental developments in the development control framework. We distinguish here between legislative instruments, policy instruments, and initiatives. Legislative instruments represent primary legislation, enacted in and by Parliament. Policy instruments include those statements that are issued under the direction of the Secretaries of State: for example, government circulars and Planning Policy Guidance Notes. Finally, our use of the term 'initiatives' is intended to indicate a more amorphous government policy which, whilst it does not carry the legal weight of the other two categories, represents an important ideological position of the government.

Legislative instruments

Legislation did not play a significant role in the Thatcherite reorientation of the development control framework in the 1980s, with the exception of the Local Government Planning and Land Act 1980. The achievements of this piece of legislation were, however, significant, particularly in establishing the necessary legislation for the Urban Development Corporations. It is also important to identify the important role played by prospective legislation in the reformulation of development control in the 1980s. The most notable example is the White Paper 'Lifting the Burden' (H.M. Goverment 1985) which did not find its way on to the Statute Book, although one cannot underestimate its importance as a statement of the government's philosophy. The 1990 Town and Country Planning Acts were consolidating in nature, in preparation for the Planning and Compensation Act 1991. The changes established by the 1991 Act relate principally to the status of development plans. This has inevitably had important implications for development control practice, and decision-making in particular. Yet even here, it is too early to ascertain whether the changes brought in by the 1991 Act will have significant implications for decision-making in development control (Gatenby and Williams 1996; MacGregor and Ross 1995; Tewdwr-Jones 1997). The 1991 Act also initiated a statutory basis for the notification of proposed developments to neighbours. This clarification of rights and responsibilities is a persistent feature of change in development control in the 1990s, and is considered further below.

Policy instruments

The increasing government use throughout the 1980s of non-statutory policy instruments, especially government circulars, as a means of realising rapid changes in the orientation of the development control framework is clearly illustrated by Thornley (1993). Circulars 22/80 'Development Control Policy and Practice' (DoE 1980) and 14/85 'Development and Employment' (DoE 1985) are widely regarded as radically reorienting development control towards a more streamlined and efficient system and provided the first point of influence for the government to eradicate inefficiency in the planning system. This is especially so in the context of removing obstacles to businesses and generating industrial development. It is without contention that the agenda was one of deregulation of the development control framework.

The introduction of changes to the General Development Order in 1988 was a central component of development control in the Thatcher years. Various amendments continued to be made to the Order. In 1995, the government issued two Orders, the Town and Country Planning (General Permitted Development) Order and the General Development Procedure Order 1995 which replaced the 1988 Order. The 1995 Orders are essentially an exercise in consolidation, and therefore as a result incorporate the changes made in the late 1980s. Nevertheless, there are a number of changes as a result of the 1995 Orders. First, where a development permitted by the Town and Country Planning General Permitted Development Order requires environmental assessment, those permitted development rights are withdrawn. This is essentially a response to European legislation.

Two other changes relating to conservation areas are also of importance. The Order inserts a new Article which enables a local planning authority to issue a direction to withdraw domestic permitted development rights within conservation areas without the need for approval by the Secretary of State. Further changes enable a local planning authority to withdraw permitted development rights relating to demolition within conservation areas. These two changes identify a greater element of local discretion, accompanied by a commensurate level of responsibility, conferred upon local authorities. This contrasts with the manner in which control was redistributed in the 1980s, in which control was first centralised and then transferred to selected private and industrial interests.

Elements of deregulation in development control continue to be identified. For example, Statutory Instrument 298 of 1995 introduced a permitted change of use under the General Development Order that allowed a change of use from use class A1 or A2 of the Use Classes Order to a mixed use for purposes of either of those classes and as a single flat from such a mixed use to either of those classes. The Town and Country Planning (Use

Classes) Amendment Order 1995 (SI 297) repealed the former classes B3 through to B7 (known as the Special Industrial Groups) which incorporated the heavy and 'neighbour-unfriendly' industries. Activities within these former categories are now included in use class B2. This is a significant change in planning terms alone, although the various other environmental regulations that often apply to these activities diminish its significance in practice. The government's rural White Paper of October 1995, 'Rural England: A Nation Committed to a Living Countryside' further proposed a new rural business use class, illustrating that there continues to exist a special concern for rural areas within the government in the 1990s (H.M. Government 1995).

Initiatives

The post-Thatcherite era has been characterised in part by a series of initiatives that have impacted upon very many areas of public policy. The example that will be expanded upon below relates to development control as a service. This has been realised principally through the Citizen's Charter initiative, launched by John Major in 1991 (H.M. Government 1991). The Charter initiative is best viewed as a vehicle for furthering the aims of efficiency and quality, which are not exclusive to it, and certainly precede it. Its application has been across the public policy field, although it has been readily adopted in and proven appropriate to planning. The use of initiatives such as the Citizen's Charter is a much softer approach to reorientating development control practice than we witnessed in the 1980s. The post-Thatcherite era does not express the aggressive attack on development control that was so readily perceived in the early 1980s. The initiatives to date have placed considerable emphasis on efficiency and on the clarification of responsibilities. This has inevitably resulted in a focus on process and procedure in development control.

It is possible to discern an increasing use of softer initiatives in the 1990s as opposed to strong and overt policy. Government planning policy in the 1990s had no identifiable 'centrepiece', a title which Thornley (1993) ascribes to Circular 22/80. This is not to say that statutory instruments have not continued to play an important role in the government's shaping of development control practice. These traditional instruments remain central to development control and therefore remain effective instruments of change. Examples are given above of their use in the deregulation of the planning system, although not on the scale of the previous Thatcher governments'. In development control, in particular, the emphasis which has been most effective in reorienting development control practice is that of procedure, rather than policy. Whereas Thornley (1993) identifies procedures as a 'second dimension' in the reorientation of the planning system, we identify procedures as the primary dimension of change in development control in the 1990s. This typifies the New Right approach to development control in this period. This point is elaborated in the following accounts of contemporary development control practice in Barrow-in-Furness, Cumbria, and in Newport, South Wales.

Development control as a public service: are you being served?

In the last few years, questions relating to quality and efficiency in development control determination have been firmly placed on the political and public agenda. The scrutiny of both local planning authorities' decisions and the legitimacy of those decisions is becoming intensified. The formerly cosy world of the planning committee is no longer a safe haven for uncertainty: every recommendation, every discussion, and every decision made by officers and councillors on every application will have to be watertight (Tewdwr-Jones 1995). The quality aspect of the development control process has been examined in detail relatively recently by the Audit Commission, which considered the delivery of the development control service (Audit Commission 1992). The Commission's review emerged partly in recognition of the need for guidance under the 'plan-led' planning system, and partly to publicise the government's commitment to effective, efficient and economic public services. Referring to the importance of procedures under the post-1991 planning system, the report states that:

> Development control is the executive arm of the planning process. It gives effect to the planning objectives of the development plan. The quality of the outcome is critically dependent on the quality of the development plan and the extent to which individual planning decisions are consistent with it.
>
> (p. 45)

The Commission was particularly concerned to assess the objectives of the development control process and examine the quantitative and qualitative performance indicators that can be used to measure the efficiency of the planning process. Given the government's desire to strive for efficient public services, questions of quality and accountability are paramount. As the report further stresses:

> Quality has become a touchstone for competitive advantage in public sector activities; in the public arena it is seeking clearer definition. Major political parties now espouse some form of quality/citizen's charter.
>
> (p. 18)

The quality issue the Commission refers to, however, is not assessed in any in-depth way. Quality in development control is referred to generally as a vital part of public service provision. But it is also possible to identify different aspects of quality within development control. The process is not only dependent on the quality of service delivery – important though it may be. It also involves assessing the quality of the policies and plans the decision-makers work to, the quality of policy implementation, and the quality of actions in the decision-settings. Quality varies across all these interrelated tasks. Any assessment of the 'quality' identified by the Audit Commission is unlikely to identify the political and judgemental conflicts that exist in any determination procedure. Measurement of this quality also masks the variety of underlying managerial and political methods of service used and concentrates purely on outcome. To focus generally on 'quality in the planning system' neglects other important issues such as needs, values and judgements in decision-making contexts – issues that cannot be easily divorced from the process.

The Department of the Environment, the Welsh Office and the Scottish Office have issued comprehensive sets of guidelines to local planning authorities on the public service ethos of the development control process, in reply to the Audit Commission's report. In 1993, the government issued a 'planning performance checklist' for development control decisions, but concentrated on local authorities in their promptness in deciding applications within eight weeks. These league tables are certainly not as sophisticated as the performance indicators called for by the Commission, and there is some dismay at the DoE's wish to develop 'a crude handling time assessment' of the development control process. In Wales, the Welsh Office's 'Development Control – A Guide to Good Practice' document aims to help local authorities improve their development control practice and increase quality in public-sector service delivery (Welsh Office 1993). The guide was produced as part of the Citizen's Charter initiative and relates to the standard of service expected of development control sections. The guide states that every citizen has the right to expect the local development control service to provide seven key qualitative indicators that can be regarded as the parameters for local authorities' planning control procedures in the future. These indicators are:

1 A recognition of the importance of the control of planning and development;
2 A comprehensive local plan coverage of the whole of each authority's area;
3 An opportunity for citizens to participate in the review and drafting of development plans;
4 A requirement for planning decisions to conform to the development plan and for departures to be adequately explained;

5 A need for local authorities to respond positively to planning enquiries, promptly;
6 A determination for the profession to minimise costs and delay in the planning system;
7 A desire for good administrative and professional practice, for decisions to be made in accordance with plans, and for any complaints to be thoroughly investigated.

Although the document does not provide a detailed repetition of the Audit Commission's report, it does place the issue in context for the delivery of the development control service.

The development control process has, as a result of government efforts, come to be seen more prominently as a public service during the 1990s. The traditionally held view of planning as a mechanism for the regulation of land and property development (Ambrose 1992) has been accompanied by one in which development control offers a service to those that use it. The rhetoric of the market has found its place in planning practice: public officials talk of clients, markets and customers. The rhetoric has been widely adopted throughout local planning authorities. Yet, notions of 'the client' have no well-developed tradition in the field of land-use planning, which has traditionally operated within the interests of a widely defined 'public interest'. This and other factors have to be considered carefully if the notion of development control as a service is to have any substance and not remain a superficial exercise. Given these difficulties, the commodification of the planning service has only been partial. In this section, we identify this process of commodification of development control through two key interrelated but separate elements: the concern with costs and efficiency and the introduction of the Citizen's Charter initiative, using two case-study examples.

Costing the development control service: the experience of Barrow-in-Furness

Of the elements of the commodification of the development control system, one of the greatest impacts has been in the study of its costs. The Department of the Environment (1994a) had published 'The Costs of Determining Planning Applications and the Development Control Service' and this had provided the clearest indication of the government's stance on the issues of costs and fees in development control. The report was also prepared at a time when there was much discussion in the professional planning press on the setting of fees.

Fees for the determination of planning applications had been introduced in 1981 following the implementation of the Local Government Planning and Land Act 1980; fees for development control are therefore comparatively recent. The level of the fees has subsequently been set in regulations issued at the national level by the Secretary of State. In the period from the introduction of the fees until 1990, the government indicated that the basis upon which the fees should be set is that local planning authorities would recover half of the costs incurred in determining planning applications. This basis no longer applies and the current intention is that 'fees should meet in full local planning authorities' costs of determining planning applications' (Department of the Environment 1994a: 3). Fees are intended to rise periodically until this objective is met. The issue of the recovery of fees is a distinctly national agenda, which is suggested by the statement that 'There has not been hitherto a preoccupation with the extent to which fees cover costs' (ibid: 13). It would be wrong to suggest that local planning authorities had not previously been concerned with the costs incurred in development control or that there were sufficient resources to finance it. Yet never before had the issue of fees on an individual and 'ring-fenced' basis been properly considered in local government. The research conducted on behalf of the Department of the Environment acknowledges the mixed reaction to the introduction of application fees, yet reports that 'fees are now fully accepted'. The report, however, continues to hint at a less than full acknowledgement of the full recovery of costs through fees.

An example of the explicit concern with fees on development control is provided by taking the example of Barrow-in-Furness Borough Council, a local planning authority situated in the north west of England. The council compares well statistically to the Department of the Environment's research findings in a number of areas. For example, the DoE's research identifies wide variation between local planning authorities in the level of fees as a proportion of total development control costs, between 14 per cent and 59 per cent, with the average at 31 per cent. The corresponding figure for Barrow is 45 per cent. This is particularly impressive, given the relatively low number of applications submitted to the authority compared to other local authorities. Similarly, in the DoE research fee income was estimated on average at 66 per cent of the costs of determining planning applications, with a median of 58 per cent. The corresponding figure for Barrow-in-Furness is 59 per cent. In terms of the breakdown of costs by function, the council aligns very much with the average as reported in the DoE research.

The DoE's research report does not identify the extent to which application fees would be required to rise in order to meet the government's objective of the total recovery of costs through fees. This is a difficult task and there are many important factors that would need to be considered. A number of these are identified in the report, in particular one of the most pressing concerns is the extent to which householder application fees may need to rise. The associated concern of the extent to which this may cause developers and householders to evade planning controls altogether remains a central issue in the objective of full cost recovery. Therefore, any consideration of the extent to which application fees may need to rise is to be speculative and crude. The research data for Barrow-in-Furness Borough Council identifies net expenditure for the year 1995/6 at £78,000 for the determination of planning applications and the provision of advice on applications. This relates to a total of 532 applications, of which 413 were fee-generating. As a crude indication and at the current level of costs, this equates to an additional £146 per application, or an additional £188 per fee-generating application, on the basis of full recovery of costs. The associated impact if this requirement were introduced would be noteworthy.

The Citizen's Charter within development control: the experience of Newport

The Prime Minister, John Major, launched the Citizen's Charter as his own personal initiative during his first administration in 1991 (H.M. Government 1991). The requirements of the Charter became effective in 1994 and the provisions of the Charter have been pervasive in their influence on both public and private services. In its planning guise, the joint English and Welsh publication 'Planning: Charter Standards' was issued at this time. This document, aimed at the public, explains the planning system and identifies levels of service that users of the planning service can expect to receive. It is intended to develop on the document 'Development Control: a Guide to Good Practice' aimed at local planning authorities and published the year previously (Welsh Office 1993). A number of the standards included in these documents read like mission statements and enshrine a customer-oriented philosophy. Those standards relating specifically to development control are more concrete and typically relate to the number of working days within which an activity should be completed. Devising quantitative indicators as measures of performance in development control has been relatively unproblematic. The increasingly procedural nature of development control lends itself to the construction of basic performance indicators. The greater difficulty has been in establishing measures of the quality of development control, not only as a process for the determination of planning applications, but also in the quality of its outcomes (Audit Commission 1992). The Audit Commission readily acknowledges that 'planning has been handicapped by an absence of any shared and explicit definition of quality in development control' (ibid: 18). Despite the repeated statements of the importance of quality in development control practice, agreement on appropriate qualitative indicators has proven elusive.

The issue of quality in development control and public services generally has performed a central role in the government's approach. There is reason, however, to doubt the sincerity of it, at least in so far as it is portrayed as a principal concern. A critical reading of the Audit Commission report suggests that quality is a residual consideration, and one to be considered currently while development control workloads are low. There is little to suggest that efficiency does not remain the overriding consideration, and that this will be to the detriment of quality concerns as development applications rise again once the economy picks up.

Our example of Newport Borough Council identifies the extent to which the Citizen's Charter initiative in planning has influenced the operation of development control within local planning authorities. 'Development and Building Control: A Charter' was published by Newport Borough Council as a response to the Welsh Office's requirements for each council to have a local planning charter prepared and adopted by 30 June 1994. The council's charter readily adopts the notions of 'customer' and 'service'. The Welsh Office's guidance on the relevant material to include in such a charter is prescriptive. It is not surprising, therefore, that Newport's local planning charter closely reflects the standard model suggested by the central government department. The Welsh Office's good practice standards are replicated in Newport's charter; for example, the period of time for responses to written communication requests (five working days) and the response to telephone requests not immediately answerable (one working day) remain identical between the two documents. There is only one standard in which the council departs from the recommended good practice: setting the target of issuing decision notices on applications within two working days, rather than the Welsh Office's recommended three working days. The local planning charter serves only a relatively limited purpose in contextualising the issues raised in the Citizen's Charter and the 1994 document. Evidence of this is found most clearly in the explanation of the prevailing local policy and advice framework for the Newport area, but there exists opportunity to include more 'local' information than is the case for the Newport charter.

It is necessary to account for how these related issues of cost, efficiency, performance and quality fit into the New Right agenda for planning. Each of these elements has proven to be an effective means of realising 'hands-off control' of the planning system. Paralleling the theme of shifting control between interests, namely in favour of private industry in the 1980s, these initiatives can be seen to be strengthening customer control through the centralisation of performance criteria. Central government has clarified the rights of customers by strongly advocating the adoption of charter standards. As identified in the above case study, there exists little variation between local charters and recommended good practice.

There has been little resistance to these initiatives at the local level, despite the presence of dissenting voices. The reasons for this are not wholly clear. It is possible that this is a consequence of a diminished capacity or will to resist, through a tighter drawing of the parameters within which the activity of development control takes place. The increasingly limiting nature of the parameters defining legitimate function and purpose of development control is a possible explanation for one of the most interesting of recent phenomena in planning control. This is the operation and utilisation of development control outside of its state-defined parameters. It is this issue of locally divergent political strategies within the development control service that is taken up in the remainder of this chapter. Before we undertake this review, however, it is first necessary to revisit the debate concerning the statutory institutionalised context for planning control.

Locally divergent political strategies in development control

Statutory planning in Britain is firmly entrenched in land-use issues; social and economic concerns, although they can be referred to, are not matters that the planning process can address directly. The courts of law and central government planning policies have indicated what land-use factors can be taken into account when determining planning issues, although little guidance is provided to decision-makers on the weight to attach to these multifarious considerations (Tewdwr-Jones 1995). This is quite deliberate on the part of the state and emphasises the discretionary role (i.e. non-prescribed nature) of British statutory planning. The issue becomes more problematic when decision-makers are faced with determining which considerations to take into account. The existence of land-use policy material considerations does not necessarily imply that planners have to make decisions in particular ways. It is quite possible for planners to ignore considerations, such as central and local state policies, and take decisions on the basis of other factors. Plans and policies provide only a broad framework, not a blueprint, and this permits decision-makers discretion to determine how different issues are to be weighed up in the decision process.

British planning is not a system of zoning: flexibility is an inherent part of the statutory planning system and neither central government nor the legal system can intervene at the local level if local development controllers play-down state policy considerations, for example. The sole legal requirement is for decisions to be reasonable. If any departures are made to established planning policies, clear justification and reasoning is required to explain the anomaly. The discretion that is available, therefore, is difficult to theorise and can cause difficulties for professional officers and local political representatives in their roles as decision actors. The 1991 Planning and Compensation Act has also affected how the development control system is operationalised; the 'plan-led' clause has pushed the planning system much further down the road of establishing a primacy for predetermined policies, over which national and local planning policies must have some influence. The legislation is founded on New Right ideology, in that it introduces a greater 'rule of law' approach than previously in requiring development controllers to normally follow the provisions of the plan (Allmendinger and Tewdwr-Jones 1997). Although an element of discretion remains, there is now a clear requirement for local decisions to be founded on established policy priorities; in essence, the post-1991 development control system is one that attempts to ensure 'certainty with flexibility' (Booth 1996; Tewdwr-Jones, forthcoming). The transferability ease of planning policy into planning control (i.e. through implementation) is a process that is too recent to be tested within practical planning settings, although some commentators have already suggested that the 'plan-led' system does not necessarily imply a more straightforward task.

With an element of flexibility still apparent and decision actors recognising the availability of some discretion, the planning system in some areas of Britain has occasionally been operated by local democratic representatives in a manner the statutory planning process was never designed to address; that is, as a political tool to implement positive discrimination in favour of the 'disadvantaged'. Britain does not implement a system of planning discrimination, unlike, say, the United States. The Code of Conduct of Britain's Royal Town Planning Institute possesses no special responsibility for 'disadvantaged' groups and insists that planning should not be operated to favour particular groups in society. Social and economic concerns, although principal issues local democratic representatives will be concerned about, cannot be used as the justification for decision-making in local planning authorities. As a result, a distinction emerges between the rationale politicians refer to when making decisions (a reflection of local socio-economic ward and authority problems) and the technical and professional factors planning officers are concerned with (physical, land-use or spatial matters). The existence of discretion, the unavailability of manuals and rule-books, is viewed by local politicians as an opportunity to consider non-statutorily defined land-use problems, notably moral questions on the effect planning decisions might have on communities. The implications of this opportunity have been for the development control process to be manipulated by local politicians (particularly within rural areas) to protect territorial advantage, to promote a nationalist or regionalist agenda against institutional (state) constraints, and to instigate a modified planning approach that, since it is not set within statutory boundaries, has been labelled 'malpractice' by central government. Although this reflects the narrowly defined nature of statutory planning, it also raises the location of the development control system within complex administrative and political processes.

Planning non-conformity in practice: recent evidence

The years since 1990 have seen a preponderance in the reporting of alleged irregularities in the development-control decision-making duties of local planning authorities, as one would expect from any increase in the scrutiny of public services. The actions and decisions of both planning officers and councillors have been placed under the spotlight and questions relating to the relationships between each group questioned.

The initial reporting of alleged irregularities within development control occurred in December 1991 when Channel Four television screened a documentary entitled *Cream Teas and Concrete*. The programme directly considered cases of alleged malpractices and irregularities in the planning decisions of North Cornwall District Council, a local authority in the south west of England. The issues raised in the programme have had serious implications for the way in which all local planning authorities conduct themselves. It was not until the summer of 1992 that the government established an official inquiry into the allegations under the control of Audrey Lees. When the final report of the inquiry was issued in November 1993, Lees found that serious malpractices had occurred in the district since the late 1980s in the carrying out of its development control functions. In particular, she stated that:

> In the past committees have not acted in a wholly consistent manner, and have been prompted to take perverse decisions, sometimes by the urging of certain councillors against officer recommendations. Such councillors have been giving priority to certain categories of planning applicant, rather than discriminating on the basis of land use, which should be the main criteria.
>
> (Department of the Environment 1993a: para. 4.04)

The inquiry had been informed that the local politicians were giving preference to families and other groups who had been resident in the local area for generations. Such groups included, 'farmers, rugby clubs, Methodists, freemasons, developers, and builders' (ibid: para. 4.01). Planning permission had been secured by the formation of 'pacts' between councillors on the planning committee. The irregularities and legitimacy associated with this particular form of decision-making was further compounded by the number of planning permissions awarded to the planning committee members themselves. Lees reported that:

> In one year 13 planning permissions were obtained by members of the planning committee or their close relatives, and I have seen councillors' applications which represent the very worst examples of sporadic development in the countryside.
>
> (ibid: para. 4.26)

The report did not find evidence of any corruption nor criticise the competence of the technical staff or planning officers. Virtually all of the recommendations concern the actions of the politicians in ignoring established national and county council planning policies. The then Minister for Planning, David Curry, remarked following publication of the report that the North Cornwall councillors were guilty of 'large-scale incompetence, inadequacy and impropriety'.

Shortly afterwards, the House of Commons Welsh Affairs Committee decided to examine the role of the planning system in rural Wales, within the context of the provision of affordable housing. Their report, 'Rural Housing', was published in 1993 following a detailed examination by the Members of Parliament into the operation of the planning and housing processes in the 'Welsh heartlands' of mid-, west and north Wales (House of Commons Welsh Affairs Committee 1993). This followed local media and community complaints over the apparent misuse of the development control system in these areas to achieve non-planning gains. In its analysis of planning in rural Wales, the Members of Parliament commented that, 'Many of the most disturbing aspects of the evidence we have received [in relation to planning] have related to the conduct of members of planning committees' (ibid: para. 56). The report quotes the evidence of the Director of Planning at Ceredigion District Council, a predominantly rural local authority, in which he remarked that the development control system in the locality had 'become personalised to the extent that the circumstances of the applicant are frequently considered to be more important than the planning merits of the application' (ibid).

A political leader of Ceredigion District Council, an area of mid-Wales where a strong opposition to the system and processes of planning exists, complained in evidence of the imposition of English policies at the Welsh local level: 'The Secretary of State has gone so far as to say that the local communities and local people should decide where and what sort of development is to take place in their areas. We are guided by the policy guidance note' (ibid: para. 23). The member then went on to justify the local authority's taking an opposing view to national guidance, preferring instead to rely on 'local knowledge' provided by the area ward representative in determining individual planning applications. Cloke (1996) has also highlighted the problems of development controllers in rural Wales in applying housing and countryside planning policies of central government, and has suggested that there exists a political reaction against the imposition of such 'Anglocentric' guidance:

> It is hard to escape the general conclusion that the degree to which statutory planning provisions, or legally significant planning policy guidance, are adhered to by local planning authorities

> depends on local political agendas and localised constructions of 'what it is possible to get away with'.
>
> (Cloke 1996: 300).

However, it should be noted that these features are also found in parts of England and Scotland too, and Cloke's comment (although made in relation to the Welsh cases) can equally be valid for non-Welsh localities.

The politicisation of land-use planning, as identified by the Welsh Select Committee, is occurring as a result of the actions and decisions of the elected members of certain authorities. As a direct consequence of the imposed rigidity of national policy – and ultimately decision – parameters, some local politicians are becoming frustrated with their inability to apply local interpretation of national guidance in local circumstances. The professional officers are conforming to the policy planning constraints, but the elected members are reluctant to follow imposed central government guidance. This reluctance to conform is, in turn, resulting in occasions when the local members reject professional advice, downplay central government planning policies, and thereby contribute to reports of misrepresenting the purpose of the development control system. Although this is part of a political reaction, it is especially difficult for a planning officer to deal with. The professional has to recognise and manage politicians' requests to implement an anti-institutional decision-making process against the imposition of central and local government policies.

Cloke (1996) recognises the parallels here between the local political configurations of the Welsh rural authorities and a report of maladministration in the operation of the development control function of North Cornwall District Council and many issues highlighted by the Members of Parliament in Wales are similar to the contents of the North Cornwall report.

Authorities in both the Welsh 'heartlands' of west, mid- and north Wales and North Cornwall reflect to some extent the same socio-political characteristics. These authorities are politically Independent, are centred on geographically marginalised rural areas, and possess broadly similar socio-economic conditions, such as rural deprivation and lack of affordable local housing. These socio-economic factors and tensions within both rural Wales and North Cornwall have undoubtedly impacted upon the operation of development control. Among the contexts for their planning-control decision-making are a number of socio-political issues that are quite distinct. These include:

- the removal of the 'safety valve' for local needs provided by local-authority rural housing and the impact of the 'right to buy' legislation associated with housing association programmes in rural areas;
- the general socio-economic changes in rural areas, including the

declining agricultural and quarrying labour force, and the effect of rural depopulation;

- the pressures provided by growth sectors such as tourism and their 'disruptive' impact on existing social structures in the countryside; and
- the consequence of in-migration into these areas of people no longer economically or socially dependent on those traditional structures.

Additional to these societal problems is a cultural phenomenon. In rural Wales, the erosion of Welsh identity and the threat to the Welsh language are compounded by a supposed 'threat' from non-local people migrating into the area and eroding both the low supply of housing stock and the distinctive cultural circumstances, although this latter point does not seem to be as apparent today as it was in the 1970s and 1980s. It remains, nevertheless, an entrenched political concern. The political representatives of both rural Wales and North Cornwall are attempting to take decisions within the development control process that meet the needs and problems of the communities they serve while territorially conserving the cultural nuances that are identified as distinctive from the rest of the country. While it is questionable whether this forms the basis of a rural local state (Cloke and Little 1990), planners in these areas are faced with a problem of local politicians refusing to enact central policies institutionalised and backed by a legal system. The decision-makers view with great suspicion the imposition of outside policies or guidance that have not been formulated at the local level. They also do not believe that policy solutions addressed by anyone other than local people can truly reflect the problems being experienced or create sets of conditions that are appropriate for their own constituencies.

Local government and local political representatives are viewed as the best actors to respond to local needs. The impact of the above tensions on the indigenous populations of these rural areas has led to a dissatisfaction with the national government's response in adopting a systematic policy to ameliorate some of these problems. And in the absence of national policies, the local development controllers have developed a local capacity for non-official planning that, while not politically strong, responds to powerful and passionate beliefs held by the wider community. These issues are not unique to rural Wales or the south west of England, but they have been the most publicised cases as they have affected the operation of the development control system in these areas.

Planning within these locations does seem to have been at variance with the rest of the country and it has received widespread publicity for its non-conformity. However, it is important to place these development-control decision processes in perspective: it is unlikely that these malpractices could be identified in the majority of local authorities in the UK. While they may prove to be the exception rather than the rule, there is a possibility that aspects of these cases (especially the reasons for non-conformity) could be identified elsewhere. There is also some uncertainty whether these problems are unique to rural areas. The qualitative evidence provided to date has only considered the operation of non-conformist development control in rural local planning authorities. The tensions between officers and members is an aspect of development control that could certainly be identified elsewhere. However, we are theorising these problems following the evidence available to us at the time, which is predominantly rural in character.

If these development-control decision-making capacities are viewed, not as a process of maladministration *per se*, but as the product of political representatives' utilising discretion to achieve discrete socio-political benefit, could other localities display similar characteristics? Localities' land-use planning non-conformity might also be the product of a local political reaction against the imposition of an institutional system considered inappropriate for local spatial circumstances. In its place, an alternative planning agenda is developed for local communities by their political representatives. Although the British planning process does not permit the existence of sub-national planning regimes, separate to the British 'state model', planning decision-makers have deliberately attempted to implement statutory processes through development control that, while not illegal or corrupt, are disobedient to the national policy process of the state. The difficult question for professional planners to reconcile is: how to act within a top-down legal land-use planning system provided at the national level while truly reflecting local democratic discourses that may have non-land-use connotations and are contrary to central government's policies? The discretionary role of the decision-maker therefore becomes one of great concern. Such individualistic dilemmas for planners are rarely defined or explained in the literature. But recent questions of what role planning should take at the local level to promote greater democratic involvement has precipitated an associated discourse on state–local activity. It has given rise to concern that planners are situated at an unhappy disunion between balancing local political capacity for discretion with national institutional and professional constraints.

Conclusions

The post-1991 planning system, the increased scrutiny of public services through the Citizen's Charter, and the move towards more open government are all likely to change the nature of planning-application decision-making in local government. The internal management of planning services and the formal and informal relationships between planners and politicians at the planning committee will also take on an increasingly important role with developers searching for any differences of opinion in order to progress successful appeals to the Secretaries of State. But the last Conservative government was concentrating too heavily on measuring the outcome of development control to the detriment of assessing the processes utilised within the service. The discretionary arrangement of planning in England and Wales is being severely tested by the enforced uniformity of procedures that is standardising both policy formulation and implementation. The standardisation-of-procedures hypothesis may be questioned by the professional officers, but the elected members are prepared to give it some sympathy. Local decision-makers are eyeing increasing monitoring of decisions by the Audit Commission and local ombudsmen, the bombardment of national policy guidance from central government, and the issuing of good-practice 'advice' as threats to both local democracy and local independent accountability.

Commentators have suggested that the centralising tendencies of the Thatcher governments toward planning issues were not as widespread in rural areas as the measures introduced in urban centres, and that the interventionist nature of central government's involvement in the planning process in the countryside was less apparent, being far more dependent on localised discretion operated by local planning authorities (Cloke 1996). While there is truth in the notion that the centralising tendencies of the government during the 1980s and 1990s appear to have a strong urban focus, it would be wrong to suggest that rural areas faced a less-reinforced regime in the policy-making process. If anything, the restrictions to local policy- and decision-making processes were just as marked in the countryside as they were in the towns and cities. The difference is rather in the form intervention took.

Central government initiated a plethora of urban policy initiatives that by-passed local government, including central grants, Urban Development Corporations and Enterprise Zones. Although these type of non-statutory planning initiatives were not apparent in rural areas, the statutory planning policy process in the countryside – the institutionalised part of the planning system – was placed under similar centralising tendencies by central government to those operating in urban centres. The form centralisation took included releasing circulars and Planning Policy Guidance Notes within which central government set out the broad planning policy framework. Local planning authorities, in both their policy-making and planning-control functions, are legally required to adhere to this 'guidance' and there can be little doubt that rural planning authorities' planning agendas were compromised by central policy planning statements to an equal extent as urban local government's.

A great deal of evidence has been put forward to suggest that rural planning policies have been different from those for urban areas, and that rural decision-makers have attempted to implement separate agendas against the wishes of central government. Two separate inquiries in North

Cornwall and rural Wales have uncovered 'malpractice' in local planning matters and non-conformity between local and national planning agendas. While it is a very attractive proposition to suggest that it is local discretion alone that has caused these differences in the operation of planning compared to the rest of the country (based on an interpretation of 'local needs'), this does not necessarily explain the reasons why a 'localised planning agenda' has emerged. The same level of discretion within the statutory planning system has been available to urban local government on planning policy and control matters and there is evidence that local planning 'malpractice' has also been occurring in urbanised areas.

It is clear from the evidence presented to both the Welsh Affairs Committee and the Lees Inquiry that the local political representatives feel passionately about 'doing something' for the communities they represent and have attempted to modify the development control process to achieve socio-economic benefits. While the British statutory planning system does not permit the consideration of non-land-use issues or the adoption of a process of positive discrimination in favour of distinct interest groups within areas, professional officers are having to mediate in the arena of discretionary conflict. Politicians are making decisions based on the socio-economic plight of their communities, while professionals are constrained in implementing technical legal and policy requirements as determined by the state and development control regulations.

As a consequence, it is possible to conceptualise the planning system as operating in distinct socio-political circumstances, according to the political ruling of each authority, the social, economic and cultural circumstances of the spatial area, and the ability of external interests to influence the decision-making processes. Development control practice in Britain should not, therefore, be regarded as a uniform administrative process, but rather as individual local political systems operating within central state boundaries. Different areas of the country will operate the development control service in different ways. In authorities where local political authority is vociferous (such as in rural Wales and the south west of England), professional officers are severely handicapped by politicians who possess clear views of how 'planning' should be defined and what it is supposed to achieve to meet local needs. No matter how many official warnings are provided by central government when official policies are ignored within local development control practice, this will not alter how these authorities take decisions. The powerful, traditionally independent mobilisation of political will in these localities against central state interference enforces a defined decision-making process that militates, perhaps not simply against a 'national' approach, but rather against any 'non-locally' defined solution. Institutional and state models of planning are rejected and in their place a modified locality-led planning system is attempted to be operated to benefit local residents' needs.

The developing notion of the public as clients or customers of planning with the associated assumption that 'the customer is always right', may have signified to the public and local politicians that the planning control system exists to provide them with what they demand, since it is central to the public service. We are not suggesting that the development of the ethic has created the pressure upon local politicians to meet through malpractice the socio-economic needs and concerns of their constituents *per se*, for these reactions are part of a wider phenomenon; but the impact the ethic may have had in some localities in the country could stand as a contributing factor.

Acknowledgements

The authors wish to express thanks to: Paul Cloke, University of Bristol; Karime Hassan, Newport County Borough Council; and Mark Robinson, formerly with Barrow-in-Furness Borough Council, for assistance and helpful comments received in drafting this chapter.

8

COUNTRYSIDE CONSERVATION AND THE NEW RIGHT

Kevin Bishop

Introduction

The Conservatives have traditionally been seen and portrayed themselves as 'the party of the countryside' (Conservative Research Department 1995: 1), enjoying electoral support from the rural shires. This rural connection, somewhat eroded in the 1980s, was rediscovered in recent years and was never more obvious than at the 1996 party conference, where the ideals of country life proved central. Douglas Hogg, then Agriculture Minister, proclaimed that Conservative values were at heart rural values. John Gummer, then Secretary of State for the Environment, expressed the hope that Tory patriotism would not be further 'clouded by urban thinking'. Whilst elements of the New Right agenda of 'liberalisation' can be witnessed in countryside planning and management (notably through continued commitment to the voluntary principle) the rhetoric does not always marry with the reality. There has been a tension between New Right ideology and traditional Tory paternalism with the 'radicalism' of the New Right agenda often tempered and reversed in the field of countryside conservation by paternal concern for the countryside and historic heritage which are more akin to the views of 'one-nation' Tories. When the Conservatives came to power in 1979 the countryside was largely unregulated in terms of planning controls over rural land uses which made it difficult for the new Government to demonstrate the deregulatory zeal associated with the New Right. In the 18 years that have elapsed since 1979 the range of regulatory controls over most countryside activities and developments has expanded: there has been an increase in the number of categories of protected areas and their spatial coverage, and the machinery of conservation governance has expanded.

A deregulated, re-regulated or newly regulated countryside?

When the first Thatcher Government came to power in 1979 they inherited a largely unregulated countryside: whilst there were strict controls over newly built development in rural areas (except for agricultural or forestry purposes) there were minimal controls over rural activities and operations (e.g. afforestation and agricultural improvements) and the land-use and landscape changes that they initiated. The Town and Country Planning Act 1947, despite its name, effectively established a system of town planning that was meant, amongst other things, to protect rural areas and agricultural land from urban encroachment. Thus the definition of development contained in the 1947 Act, and all successive Acts, effectively excludes the use of land and buildings for agriculture and forestry; hence there is no need to obtain planning permission for most agricultural or forestry operations/developments. From the point of view of rural land use it has been argued by Newby (1980: 288) that planning has been:

> virtually a by-product of a system designed to cope with urban growth, partly because the countryside was regarded as a bucolic backdrop to life in urban areas and partly because the idea of a planned countryside was, to influential public opinion, anathema . . . only in the last decade or so, therefore, has rural planning not proceeded by default.

The countryside that we see today has been shaped more by the powerful economic elements of agricultural policy than by the regulatory powers of the town and country planning system. Beyond the town or village envelope the vast bulk of rural land was not covered by development plans. It was simply designated as 'white land': land where there was a presumption against commercial or housing development. Even with the advent of structure plans in the late 1960s countryside policies remained partial and general.

The past half-century and, in particular, the last 18 years have witnessed profound changes in the countryside as the agriculture industry pursued policies that until recently were aimed at maximising production. With mounting evidence of the environmental impact of modern farming and forestry practices pressures have increased for formal environmental safeguards to protect the countryside. This debate has to some extent been portrayed as between New Right advocates of *laissez-faire* voluntarism – conservation through persuasion – and advocates of control who have argued for an extension of planning and other forms of regulation over all rural land uses. Whilst such a distinction is simplistic and largely counter-productive, as controls can only prevent removal or alteration but do not ensure sympathetic management, it does help demonstrate the context within which successive Conservative Governments have operated and the internal tensions between different factions of the Conservative Party.

The implementation of Lord Porchester's inquiry into 'Land Use Change in Exmoor' provided the first litmus test for the new administration's policy on countryside conservation. The 1974–9 Labour Government had intended, through its Countryside Bill, to introduce controls over agricultural practices in environmentally sensitive areas. In particular, the Bill contained clauses that would allow Ministers to designate specific areas of open moor or heath within which National Park authorities would have been able to make moorland conservation orders to prevent agricultural 'improvements' detrimental to the environment. The in-coming Conservative Government rejected moorland conservation orders, planning controls or any other form of compulsion to enforce conservation policies. Instead, it made the voluntary approach the centre-piece of the Wildlife and Countryside Act 1981. The voluntary approach was seen, in part, as an expression of New Right ideology – a defence of the autonomy of private property and minimal state intervention. It also reflected the Conservatives' traditional defence of the ideology of land ownership and practical political considerations – rural landowners and farmers formed a significant part of their constituent political power base. Yet, there were some contradictions: while Ministers were at pains to stress that their approach enshrined the voluntary principle, in fact the Act introduced a power of last resort to protect Sites of Special Scientific Interest (SSSIs), but not National Parks.

Whilst rural areas have not been exempt from the efforts to streamline the planning system and reduce planning controls the impact of such measures has been less pronounced in the countryside. As already noted, the scope to demonstrate deregulatory zeal was curtailed by both the comparative lack of regulatory controls in rural, as compared to urban, areas and mounting pressure for increased regulation to protect the countryside from the environmental destruction wrought by modern agricultural and forestry practices. In the eighteen years that New Right ideology has helped shape Government policy the range and significance of countryside regulation has increased rather than decreased. Specific attempts to deregulate rural planning have proved more contentious than reforms to urban planning and were often ill-fated with draft circulars being withdrawn or re-drafted and promotion of private development often tempered by accompanying guidance stating the need to conserve the countryside.

Advocacy of private development was provided in the form of a series of White Papers which outlined proposals for reducing planning control, simplifying procedures, improving efficiency and speed in decision-making, and promoting a positive approach to new development. Planning authorities, rural and urban, have been urged to take a much more sympathetic and accommodating stance towards new development. The White Paper 'Lifting the Burden' (DoE 1985d) noted that the planning system 'imposes costs on the economy and constraints on enterprise that are not always justified by any real public benefit in the individual case' (para. 3.1) and that there was therefore a need to 'simplify the system and improve its efficiency and to accept a presumption in favour of development' (para. 3.4). This new orientation was reflected in a series of Government circulars. For example, Circular 22/80 'Development Control – Policy and Practice' (DoE 1980) urged local authorities to pay a greater regard to time and efficiency and always to grant planning permission unless there were sound and clear-cut reasons for refusal. It emphasised the need for new forms of rural development and for planning policies to be less restrictive. It urged local authorities to grant planning permission for small-scale commercial and industrial activities when they were proposed for rural areas. Circular 16/84 (DoE and Welsh Office 1984b) continued this theme of rural economic development, stressing the need for new enterprises in the countryside and suggesting that 'many small-scale buildings can be fitted into rural areas without causing unacceptable disturbance'. The same circular urged local authorities to be more flexible and responsive to employment-generating proposals and not to be constrained by development plan policies. The theme of diversifying the rural economy was further developed by Circular 16/87 (DoE and Welsh Office 1987a) and PPG 7 'Rural Enterprise and Development' (DoE and Welsh Office 1988b) which superseded Circular 16/87. Yet this advocacy of private development has always been tempered by a more traditional, paternalistic concern for the countryside. Thus Circular 22/80 reminds local planning authorities of the need to protect landscapes and good-quality agricultural land and makes it clear that 'The Government's concern for positive attitudes and efficiency in development control does not mean that their commitment to conservation is in any way weakened' (DoE and Welsh Office 1980, para. 4).

Throughout the 1980s, the need to be more responsive to private developers was reinforced through the appeal system with local authorities facing a reversal of their decisions and the threat of having to pay costs. The results of this policy (increased development in certain rural areas) caused considerable unease in affected rural areas. For many villages in the south of England, the late 1980s were characterised by conflict between the Government and their own backbench members of parliament and Conservative voters. These residents were not prepared to accept the logic of a freer market for development based on ownership rights. In the public eye it symbolically came to a head with the screening by the satirical television show *Spitting Image* of Nicholas Ridley (then Secretary of State for the Environment) driving a bulldozer through Trumpton (a fictional village from a children's programme). The unprotected commuter-belt countryside (especially in the south) was the ground over which the battle between pro-market, development policies and conservation interests was symbolically fought.

One impact of this continued advocacy of private development and the need to diversify the rural economy has been to change the policy content of local plans. As Elson *et al.* (1995) note, penetration of rural economic policies into local plans is high: the local plans for their 32 case-study areas contained 238 rural economy policies – an average of over 8 policies per plan. Most of these policies related to the re-use and adaptation of existing buildings and the scale of acceptable development. From the evidence of existing research (Elson *et al.* 1995; Marsden *et al.* 1993; Watkins and Winter 1988) it would seem that local planning authorities have responded by allowing the re-use of existing buildings (including surplus farm buildings, large country houses and redundant institutional complexes) rather than allowing new development *per se*. This in turn has increased access to the countryside for non-agricultural capital (Marsden *et al.* 1993; Marsden and Murdoch 1994).

Specific attempts to deregulate and relax the constraints on development in rural areas have proved to be more contentious and often ill-fated. In 1981 a revision to the General Development Order (Statutory Instruments 245 and 266/81) developed the theme of relaxing planning controls set by the 1979 White Paper by allowing householders to extend their houses to a greater extent than was previously allowed without the need for planning permission. It allowed larger extensions for industrial premises and change of use from light industrial use to warehouse use and vice versa, for small units. However, the significance of these changes for rural areas was minimal and, in addition, the relaxation did not apply in conservation areas, National Parks and Areas of Outstanding Natural Beauty (AONBs). Attempts to relax the constraints to development posed by Green Belts and to reduce their geographical coverage proved more contentious. A draft circular to this effect aroused staunch resistance from backbench, mainly shire MPs, and rural preservation groups. This resistance, together with an inquiry by the House of Commons Environment Select Committee forced the Government to abandon its proposed reforms. In the approved Circular 14/84 the permanence of Green Belts is reiterated and their role in safeguarding the countryside referred to (DoE 1984).

In 1986, faced with declining agricultural incomes and a politically unacceptable level of agricultural surpluses, the Government established an inter-departmental working party on Alternative Land Use and the Rural Economy (ALURE). Rumours at the time suggested that there were pressures within the Cabinet for a radical restructuring of planning in the countryside with the aim of stimulating the rural economy by relaxing controls over rural business development. Such *laissez-faire* sentiments were to the fore in 1987 when the Department of the Environment issued a draft circular entitled 'Development Involving Agricultural Land' (DoE 1987a). This circular suggested that development proposals should only be referred to the Ministry of Agriculture, Fisheries and Food (MAFF) if they involved the loss of grades 1 and 2 agricultural land and were of 20 ha or more. The draft circular effectively removed the presumption that agricultural use of rural land should be paramount for planning purposes, and replaced it with a combination of concerns relating to agriculture, environment and rural economic revival:

> The agricultural quality of land and the need to control the rate at which land is taken for development are among the factors to be included in that assessment. At the same time, full regard must be had both to the need to promote economic activity that provides jobs, including the contribution of small firms, and to the need to protect Green Belts, National Parks, Areas of Outstanding Natural Beauty, and other areas of good countryside.
>
> (DoE 1987a, para. 3)

These proposals were seen by many as a serious erosion of rural planning controls, especially when viewed in parallel with an earlier consultation paper on 'The Future of Development Plans' (DoE and Welsh Office 1986a) which suggested a weakening of the strategic planning system in Britain. Local authority planning policies depended on the protection of agricultural land as their baseline (even within landscape designations such as Areas of Outstanding Natural Beauty), the reduction in MAFF input to planning decisions was considered to leave the way open for development interests to mount increasingly successful attacks on the countryside. The proposals led to headlines such as 'Is Mr Jopling [then Secretary of State for Agriculture] turning plough shares into time-shares' (*Daily Telegraph*, 13 February 1987). The Government, in the form of William Waldegrave (then Environment Minister), dismissed the fears that the new circular would lead to a rural housing bonanza as 'arrant nonsense . . . for the first time in the history of the planning system environmental concerns would be given equal priority with development' (Clover 1987). In its final version the circular made a number of concessions. MAFF was now to be consulted about development on grade 3a land, and attention was directed towards the prospects for re-use of urban land. The phrase 'other areas of good countryside' was replaced by a commitment to 'protect the countryside for its own sake', rather than simply its agricultural value. Despite ambiguity over the meaning of the phrase – does it refer to the protection of parts of the countryside already designated because of their landscape or heritage value, or to all countryside? – the final circular seemed to strengthen rather than weaken the ability of local planning authorities to refuse unwanted applications for development on greenfield sites.

The changes to the Use Classes Order introduced in 1987 in response to the 1985 White Paper 'Lifting the Burden' were described at the time as 'the biggest change to the planning system since the public participation and conservation measures of the early 1970s' (Home 1987: v). The impact of the reform of the Use Classes Order on the countryside is unclear and un-researched. Home (1987) identifies the leisure and assembly use class as having the greatest potential impact upon the countryside and coastal areas. The new class integrates two previous ones and does not distinguish between 'sport', 'recreation', 'assembly' and 'leisure'. Combined with the freedom that the General Development Order (GDO) offers to activities such as camping and caravanning, Home believes that the new Use Classes Order 'appears to offer freedom for the owners of extensive facilities such as golf courses or football grounds to turn their sites into whatever recreation uses they think fit (e.g. amusement or theme parks) . . . A farmer who establishes a recreational use on part of his land such as seasonal camping can now enjoy the freedom of this class from planning control' (Home 1987: 70–1). This argument that the new Use Classes Order confers upon the leisure industry the same freedom from planning control as agriculture has yet to be proved in practice.

The White Paper 'Releasing Enterprise' (DoE 1988a), unlike the previous three White Papers which had all focused on urban areas, promoted a radical deregulation of rural planning in order to aid diversification of the rural economy and represents a turning point in terms of New Right deregulatory zeal as applied to the countryside. The thinking contained in 'Releasing Enterprise' was developed in a consultation paper on 'Permitted Use Rights in the Countryside' (DoE 1989) which outlined proposals to relax the General Development Order in order to encourage farm diversification:

> The objective of extending permitted development rights under the GDO to a range of recreational and other activities would be to ease the burden of planning controls on farmers and others who seek alternative uses for surplus agricultural land and buildings by diversifying their activities without compromising the ability of the planning system to protect the rural environment.
>
> (DoE 1989, para. 6)

These proposals would have 'heralded widespread changes in rural land use, and would have marked a significant step in the redefinition of the countryside' (Marsden *et al.* 1993: 119). The consultation paper contained proposals to allow agricultural buildings over five years old to be used for equestrian activities, educational purposes related to agriculture or the countryside, display and sale of locally produced goods, outdoor sport or recreation, and the sale of food and drink to visiting members of the public. Not surprisingly the proposals provoked widespread outcry and they were formally dropped by the new Secretary of State for the Environment, Chris Patten, shortly after he was appointed.

The replacement of Nicholas Ridley, an outspoken free-marketeer, with Chris Patten, an acknowledged 'wet', and abandonment of the proposals to extend permitted development rights marked a turning point in rural planning. Later in the year the Department of the Environment and the Welsh Office issued a draft revised version of the PPG on Rural Enterprise and Development. The title of the new PPG was 'The Countryside and the Rural Economy' and this change of title was indicative of a change in stance with the Department of the Environment stating that 'While the Government has no present plans to extend planning controls to all farming activities, it is ready to introduce new closely targeted controls where this is necessary to deal with specific problems' (DoE and Welsh Office 1992f, para. 22).

In terms of countryside conservation there appears to be an important difference between the early years of the Thatcher administration and the advent of 'Majorism'. Between 1990 and 1997 the New Right had little influence on countryside policy and planning as successive Acts of Parliament, White Papers, PPGs and Ministerial Statements emphasised the new role of planning in the achievement of sustainable development. In terms of countryside planning, the reforms and policy development of the early 1990s more than reversed the anti-conservation rhetoric of the early 1980s. The Town and Country Planning Act 1990 signified the beginning of 'mandatory greening' by requiring all development plans to include policies in respect of the improvement of the physical environment and the conservation of the natural beauty and amenity of land. The Planning and Compensation Act 1991 made widespread reforms to the planning system and was described by various commentators, during its passage through Parliament, as one of the most important pieces of environmental legislation in the past twenty years. The 1991 Act introduced a legal requirement for all district authorities to prepare district-wide local plans for the whole of their administrative areas and increased the weight given to development plan policies for decision-making purposes (the so-called Section 54A requirement). The combined effect of the these two pieces of legislation is to bring all rural areas under the local development plan making process, often for the first time, and require local planning authorities to develop a policy framework for countryside conservation. This represents a significant extension of state involvement in countryside planning, especially when considering that as of the middle of 1989 only 66 out of 333 non-metropolitan district councils in England and Wales had local plans on deposit or adopted which fully covered their areas (H.M. Government 1990: 84).

The tone of Government guidance on countryside planning has also changed and this is reflected in the change in title of PPG 7 from 'Rural Enterprise and Development (DoE and Welsh Office 1988b) to 'The Countryside – Environmental Quality and Economic and Social Development' (DoE 1997a). A good indicator of the significance of this policy change and the ensuing extension of state involvement in countryside planning is the treatment of design issues in rural areas. Such issues did not figure prominently in the pre-1997 versions of PPG 7: planner involvement in rural design was certainly not promoted. Yet the latest version of PPG 7, published shortly before the 1997 election, endorses two techniques pioneered by the Countryside Commission to promote good design in rural areas – Countryside Design Summaries and Village Design Statements – and backs the 'Countryside Character' initiative (Countryside Commission and English Nature 1997).

Although the talk may have been about 'Lifting the Burden' through deregulation there has been a significant increase in the range of controls over rural land uses during the last eighteen years. Whilst resisting pressure from amenity and conservation interests (Council for the Protection of Rural England 1990) to extend planning controls over a variety of farming activities the Government demonstrated a willingness to introduce operationally specific and/or geographically based controls. The General Development Order was amended in 1986 to limit excavations on farmland to legitimate farming operations (e.g. fish-farming) rather than for the sale of minerals. Similar restrictions were imposed on the tipping of off-farm materials on agricultural land. In 1988 further restrictions were imposed on the building of livestock units – they are no longer automatically permitted within 400 metres of a non-agricultural building. In 1992 the permitted development rights of agricultural holdings of less than 5 ha were restricted.

More significant has been the development of so-called 'pseudo planning controls' (Bishop and Phillips 1993): controls that are implemented under planning legislation but do not question the principle of development (prior notification systems, for example); and/or, land-use regulations introduced under non-planning legislation (for example, limestone pavement orders, nature conservation orders and hedgerow management orders). The range of pseudo planning controls has increased considerably since 1980 as the Government illustrated a willingness to introduce specific restrictions on farming activities but resisted calls for a general extension of planning controls.

The notification procedure, first introduced in 1950 in areas of special landscape significance under the Landscape Areas Special Development Order (LASDO), provided local planning authorities in (and near) parts of the Lake District, Peak District and Snowdonia National Parks with a discretionary control over the siting, design and external appearance of farm and forestry developments subject to permitted development rights. This system was extended in 1986 and 1989 to cover all National Parks in England and Wales and the Broads Authority, respectively. More recently, changes to the GDO which came into force in January 1992 have extended this system of pseudo planning control to agricultural and forestry buildings constructed throughout England and Wales. Under the amended GDO, permitted development cannot be exercised unless the farmer/developer has applied to the local planning authority for a determination as to whether their prior approval will be required for certain details: siting, design and external appearance of farm and forestry buildings, roads, waste disposal facilities exceeding 0.5 ha, and fish tanks. The 'determination' procedure does not impose full planning controls – the principle of development is not relevant – but it does signify an important extension of rural regulation.

The agricultural grant notification system introduced to all National Parks by the Conservative Government in 1980 is another form of pseudo planning control. Under this system, which was designed to halt the destruction of sensitive landscapes and habitats by state-subsidised agricultural improvements, farmers seeking a grant for agricultural operations within a National Park were obliged to give advance notification of their intentions to the appropriate National Park Authority thus providing the Authority with an opportunity to suggest modifications to potentially damaging proposals.

The Wildlife and Countryside Act 1981 ushered in two other examples of a closely targeted pseudo planning control. Section 34 of the Act makes provision for the protection of limestone pavements and Section 29 provides the Secretary of State with the power to make a Nature Conservation Order for the purposes of ensuring the survival of any kind of plant or animal. Both are seen as last-resort powers and have been rarely used.

In one of its last acts whilst still in power, the Conservative Government, responding to figures from the Institute for Terrestrial Ecology that illustrated continued and accelerated hedgerow loss (Barr *et al.* 1991), provided the legislative framework for a system of Hedgerow Management Orders. The regulations make it an offence to remove a hedge that is over 30 years old and 20 metres long without local authority consent. If a hedge is unlawfully destroyed, local authorities can insist on a replacement.

One or two or more countrysides?

During the last eighteen years there has been an increase in the area and number of protected areas (an area of land and/or sea especially dedicated to the protection and management of scenic, wildlife, heritage and/or other environmental values) (Bishop *et al.* 1995). Since 1979 fifteen new categories of protected area have been established in the UK (see Table 8.1) –

Table 8.1 Categories of protected area established since 1979

Protected area	*Description*	*Date of origin*
Area of Special Protection	Designated by the Secretary of State for the Environment under the Wildlife and Countryside Act 1981 to provide protection for individual species of birds under threat from human activity	1981
Broads Authority	The Norfolk and Suffolk Broads Act 1988 provided for the designation of the Broads which is generally regarded as part of the National Park family despite having its own unique legislative basis	1988
Environmentally Sensitive Area	Designated by the Agricultural Minister under the Agriculture Act 1986 to provide financial incentives for environmentally beneficial farming methods	1986
European Marine Site	Identified under the Conservation (Natural Habitats etc.) Regulations 1994 as a Marine Site of Community Interest	1994
Forest Nature Reserve	Non-statutory reserve established by the Forestry Commission for the very best conservation sites under their management	1988
Marine Consultation Area	A non-statutory mechanism devised by the Nature Conservancy Council as a means of notifying sites of important conservation significance to potential developers and their regulators	1985
Marine Nature Reserve	Designated by the Secretary of State under the Wildlife and Countryside Act 1981 for conservation or scientific and educational purposes; MNRs can extend up to three miles offshore	1981
National Scenic Area	Set up under the Town and Country Planning (Scotland) Act 1972 to provide for the special consideration to be given to designated areas in planning matters	1986
Natural Heritage Area	Set up under the Natural Heritage (Scotland) Act 1991 for sites of outstanding value for the Scottish natural heritage	1991

Nitrate Sensitive Area	Designated by the Secretary of State using powers in the Water Resources Act 1991 to provide for voluntary participation by farmers in schemes designed to reduce nitrate loads	1991
Regional Park	An extensive area of land, part of which is devoted to recreation, created by the Countryside (Scotland) Act 1981	1981
Site of Community Importance	Potential Special Area of Conservation protected under the Conservation (Natural Habitats etc.) Regulations 1994; marine SCIs are known as European Marine Sites	1992
Special Area of Conservation	To be designated under the Conservation (Natural Habitats etc.) Regulations 1994 to protect listed habitats and habitats of listed species; SACs are to be selected from lists of SCIs submitted to the European Commission and, together with SPAs, will form the Natura 2000 series	1992
Special Protection Area	Designated under the EC Birds Directive to protect the habitat of specified species of bird	1979
Tir Cymen	Farmland stewardship scheme established by the Countryside Council for Wales using experimental powers in the Countryside Act 1968	1992
Water Protection Zone	Designated by the Secretary of State under the Water Resources Act 1991 to provide measures to prevent pollution of specified stretches of water	1991

an effective doubling in the number of categories of protected area. Almost half of these new categories have their origins in Brussels rather than Whitehall, a testimony to the growing influence of European Union membership on countryside conservation (see below). The area of land covered by protective designations has also increased. The area of SSSIs (Sites of Special Scientific Interest) in Great Britain increased by over 30 per cent between 1984 and 1995 from 1.4 million ha to 2 million ha (DoE 1996). The area of land covered by National and Local Nature Reserves increased from around 90,000 ha in 1984 to 200,000 ha in 1995 and over the same period the area covered by internationally designated Ramsar sites increased to 300,000 ha from 69,000 ha (DoE 1996). The willingness to develop and designate new protected areas to tackle environmental problems as they arise has led to what many commentators have called a 'differentiated' or two-tier countryside: areas of national or international importance have been excluded from the relaxation of planning controls and even granted extended regulatory safeguards, whilst in the rest of the countryside a more market-oriented and less community responsive planning system has evolved (Cloke and McLaughlin 1989; Lowe and Flynn 1989).

It has been suggested that 'We no longer have a planning system but three systems' (Adam Smith Institute 1983; Thornley 1981 and 1986). Areas of high environmental value (National Parks, AONBs, SSSIs) have not just been exempted from the relaxation of planning controls but are seen as important extensions to their powers of control. In the wider countryside the planning system has remained largely intact with the extension of regulation through some closely targeted pseudo planning controls while the overall emphasis has been on the facilitation of economic development with due regard to countryside conservation. In urban areas (notably Enterprise Zones, Simplified Planning Zones and areas covered by Urban Development Corporations) planning constraints have been streamlined and weakened, to allow market forces to prevail.

The Department of the Environment itself has expounded the concept of a differentiated countryside (DoE and Welsh Office 1992f). In areas statutorily defined for their landscape or wildlife qualities (National Parks, AONBs, SSSIs etc.) policies of restraint are expected to prevail. In some areas additional statutory planning controls or procedures apply, for example, through tighter controls on permitted development, and planning policies and development control decisions should 'sustain or further the purposes of that designation' (para. 3.1). The most productive agricultural land is also given some protection. Almost 30 per cent of agricultural land in England and Wales is of grades 1, 2 or 3a and local authorities are advised to give 'considerable weight' to its protection 'as a national resource for the future' (para. 2.5). For other areas of the countryside a different approach was advocated, relying much more on the facilitation of development on the grounds that: 'Maintaining a healthy rural economy is one of the best ways of protecting and improving the countryside' (para. 1.6).

Whilst it is true to state that 'different' planning regimes have operated in different parts of the country the impact was primarily urban and it is easy to overstate the practical impact of the so-called 'differentiated countryside'. Statutory designations, especially those aimed at protecting our landscape, have traditionally relied upon stricter control regimes than those operating in the rest of the countryside. In the wider countryside there is evidence to suggest that many local planning authorities have responded to deregulation rhetoric by relying on a multiplicity of local countryside designations ('Areas of Great Landscape Value', 'Special Landscape Areas', 'Special Countryside Protection Areas' etc.) to provide additional safeguards for development control purposes. Thus the distinction between statutory designations and the wider countryside becomes blurred. Somerset County Council's 'Special Landscape Areas' designation, as defined in the Structure Plan, covers approximately 60 per cent of the county and within individual districts up to 65 per cent of the land area is covered by one or more non-statutory designation. The Country Landowners Association (1995) and Rural Development Commission (1995) have both expressed concern that the proportion of countryside without additional planning restrictions is being eroded by a proliferation of non-statutory designations.

In recent years the Government has signalled a move towards considering the countryside as a whole. The White Paper 'Rural England' (DoE and MAFF 1995c) acknowledges that 'designating special areas is not, on its own, an adequate mechanism for conserving the quality of landscape and the abundance of wildlife which we all want to see. We can no longer afford to view designated areas in isolation from the rest of the countryside' (p. 105). This new phase in conservation policy was carried forward in the latest version of PPG 7 (DoE 1997a) which states that: 'The priority now is to find new ways of enriching the quality of the whole countryside whilst accommodating appropriate development, in order to complement the protection which designations offer' (para. 2.14).

Market mechanisms or market forces?

New Right enthusiasm for market forces was witnessed in the countryside but often tempered by EU membership (see below) and external lobbies. It is generally acknowledged that the reforms to the planning system, rural and urban, in the early 1980s were concerned with making it more responsive to the market but the most profound influence on the shape of the British countryside in the last eighteen years has been the Common Agricultural Policy (CAP). The subsidies, guaranteed markets and state intervention associated with this pan-European policy jarred with New Right enthusiasm for market forces and limited state intervention and in the mid-1980s the Prime Minister was credited with the belief that 'an attack on farmers' privileges' is long overdue. The first change came suddenly in 1984 and emanated, not surprisingly, from Brussels rather than Whitehall. The abrupt introduction of milk quotas signified a crisis in European agricultural policy and the dawn of an era of continuing policy reform. Increasingly the prevailing concern was to cut food surpluses and the costs of the CAP. Since the mid-1980s, Government policy, tempered by membership of the EU, has been to reduce and redirect state support for agriculture. New support schemes have been developed that aim to curb production, diversify the rural economy and conserve the countryside. The Government's approach has largely been conditioned by their enthusiasm for market forces and desire to reduce state intervention. For example, MAFF has emphasised that farmers need to adopt new scientific and technological advances so as to compete effectively in the market place and also highlighted an 'opportunity to create a market place for environmental goods' (MAFF 1991: 3). The Government's approach was influenced by bodies such as the Institute for Economic Affairs which published a paper (Howarth 1985) arguing that a free agricultural market would be a successful and low-cost way of providing food for Britain and by the New Zealand experience of scrapping all agricultural subsidies (Willis 1988).

As with countryside planning policy, the appointment of Chris Patten as Secretary of State for the Environment in 1989 marked an important change in the Conservative approach to environmental policy. The pursuit of unadulterated market forces (for example, Nicholas Ridley's proposals to privatise nature reserves), were replaced by enthusiasm for market mechanisms for environmental protection and enhancement. Patten's specialist adviser, Professor David Pearce, advocated a market-based approach to sustainable development with consumers and industry given clear signals about the costs which will be imposed upon society by given levels of pollution (Pearce *et al.* 1989). Pearce's ideas were readily accepted by Patten who stated that market-based instruments offered an 'efficient and flexible response to environmental concerns' (Patten 1990). The Government's commitment to the so-called market-based approach to the environment was outlined in the White Paper on the Environment (H. M. Government 1990): 'In the Government's view, market mechanisms offer the prospect of a more efficient and flexible response to environmental issues, both old and new' (p. 14). This enthusiasm, coupled with the shortcomings of Environmentally Sensitive Areas (ESAs) (see Bishop and Phillips 1993), has led to the development of a market approach to conservation.

The idea of landowners and farmers selling an environmental land management service was contained in the Country Landowners Association (CLA) report on 'Enterprise in the Rural Environment' (Greenwell 1989). The Greenwell report responded to the challenge of the Government's ALURE document (1987) which had enjoined farmers and landowners to be more enterprising in the marketing of a whole range of goods and services from their land, not only food. It envisaged landowners and farmers acting as independent businessmen, taking the initiative to draw up an environmental plan for all or part of their land. The decision to prepare such a plan would be voluntary and not as a result of a designation imposed by Government. The plan would be negotiated, and a contract drawn up, between the owner/occupier concerned and central or local Government. An owner/occupier might also negotiate with other public or private organisations. The CLA called this approach the Environmental Land Management Scheme (ELMS). At the national level, the market approach to countryside conservation has been piloted through schemes such as Countryside Stewardship in England and Tir Cymen in Wales. The principle behind such schemes is the creation of a market in environmental and related recreational services which can be 'acquired' from farmers and landowners. Incentives are offered from the public sector to farmers and landowners to manage their land according to certain prescriptions. The schemes allow farmers to identify relevant environmental services and goods which they can provide and the opportunity to market these and promote their role as custodians and managers of the countryside. Unlike ESAs or the compensation arrangements under the Wildlife and Countryside Act 1981, the public-sector 'buyers' of environmental and recreational services is not obliged to come to an agreement; they can pick and choose in order to get the best value for money. Discretion, therefore, is an important part of the concept. Payment is not for the process but for the product. Such an arrangement mirrors the way payments are made for most other agricultural products; it provides an incentive to produce the environmental product at the lowest cost and it encourages managerial and entrepreneurial flair.

Further insights into the Conservative Government's emphasis on moving agriculture towards the market can be found in the UK Sustainable Development Strategy (H. M. Government 1994a), the report of the Minister's Review Group on CAP Reform (MAFF CAP Review Group 1995, 1995a) and the White Paper 'Rural England' (DoE and MAFF 1995c). The objectives for agricultural policy identified in the UK Sustainable Development Strategy reflect the then Government's belief that sustainable agricultural practices will be encouraged through reforms aimed at deregulation and exposure of farming to the rigours of the market. The report of the CAP Review Group argues that a single policy is no longer appropriate to pursue separate agricultural, social and economic objectives. It concludes that further CAP reform is needed 'towards the market via reductions in end-price and other production related support' and a removal of supply controls (MAFF CAP Review Group 1995: 6). The principal aim of policy must be 'to encourage farmers to produce according to market demands, resulting in more viable businesses and a more efficient farm structure while at the same time dealing with any problems of market failure' (MAFF CAP Review Group 1995: 6; 1995a). The conclusions of the Review Group were broadly reflected in the White Paper 'Rural England'. This argues that the 'goal of safeguarding and enhancing the rural environment should be at the heart of a reformed CAP' (DoE and MAFF 1995c: 53), a goal which sits somewhat uneasily with the White Paper's vision of an 'Efficient, prosperous and outward-looking agricultural industry, able to compete in increasingly open world markets and paying due regard to the environment' (ibid). Whilst the emphasis of Government policy was on reducing state support and exposing UK agriculture to the rigours of world markets and meeting environmental objectives through a combination of 'buying' environmental goods through direct payments, advice and, as a last resort, regulation, the practical inability to force radical change in the CAP meant that agriculture remained heavily subsidised rather than exposed to the ravages of market forces.

The pursuit of market forces was more successfully achieved in terms of forestry policy. The role of the state has been curtailed through the rolling back of Forestry Commission activities and the private sector encouraged by the setting of ambitious planting targets. Up to 1988 the Thatcher administration presided over a boom in private-sector forestry fuelled by generous tax breaks, giving free rein to market forces (Tompkins 1989) which pushed conifer plantations higher up the hills, further north and on to ever more remote and poor-quality agricultural land which was often of high conservation value.

A privatised or commoditisation countryside?

The most widely recognised translation of New Right ideology into practical political strategy has been the pursuit of privatisation policies which have sought a general reduction in or redefinition of the role of the state and a shift in responsibilities for production and consumption into the private sector. Heald (1984) and Peacock (1984) suggest four categories or aspects of privatisation:

1. Privatisation in the form of charging for a service – services are charged for rather than funded from general taxation
2. Privatisation through 'contracting out' – services traditionally provided by the public sector are put out to tender in the private sector
3. Privatisation through liberalisation/deregulation of the market place – removal or reduction of constraints upon the operation of the free market
4. Privatisation in the form of denationalisation – transfer of industries or utilities partially or wholly in the public sector to the private sector.

As discussed in previous sections, the theme of liberalisation/deregulation was pursued in terms of countryside-planning, agriculture and forestry policies but to differing degrees and with different impacts. The theme of making the user pay and contracting out have also been pursued in terms of countryside conservation and often resulted in new forms of public–private partnership rather than full-blown private-sector provision. The establishment of the Operation Groundwork experiment in 1981 was one of the first examples of the change in emphasis associated with the New Right. The Countryside Commission had during the 1970s pioneered the idea of countryside management based on local authority work teams and project officers. The Commission planned to launch a major experiment run by the local authorities in the management of the urban fringe in the St Helens and Knowsley area. The Secretary of State for the Environment (then Michael Heseltine) welcomed the idea but 'suggested that an environmental trust might be appropriate and stated that, in the event of an acceptable mechanism being found, the project would receive the full backing of the Department and his personal support' (cited in: Countryside Commission 1981: 19). Operation Groundwork was subsequently established as a trust with a specific aim of involving the private sector and levering in private finance which was meant to replace public finance over time. The decision to proceed by the trust route was 'a clear indication that the tide was moving away from local government leadership in the countryside field' (Phillips 1993: 69–70). The development of landholder contracting whereby private landowners/occupiers carry out conservation or access work on their land in return for grant payments also gained credence during the 1980s and early 1990s particularly following the demise of the Community Programme which effectively resulted in the loss of some 45,000 person-days per year in practical conservation (O'Riordan 1989). The notion of the 'user pays' also dovetailed with Conservative concepts of 'active citizenship' and led to a plethora of policies and initiatives which encouraged people to take action to sustain their local countryside: for example, various Adopt-a-Path and Pocket Park initiatives. The changing role of local authorities away from direct provision and action to enablement has also contributed to developments in this field.

The selling of industries either totally or partly owned by the state to the private sector was closely associated with the Thatcher administration and its New Right tendency (Bell and Cloke 1989) and was probably the most widely identified manifestation of privatisation tendencies. As with most other aspects of New Right ideology as applied to the countryside, attempts to privatise state industries or utilities and land holdings have been tempered or aborted because of the efforts of conservation pressure groups (often working closely with statutory agencies). The privatisation of the Regional Water Authorities was the first state 'sell-off' to significantly affect conservation interests. There was concern that the newly privatised water companies would manage their extensive rural land holdings with a view to maximising revenue rather than conservation value. Following lobbying by the Countryside Commission and leading pressure groups (such as the Council for the Protection of Rural England) the House of Lords, against the Government's wishes, introduced protection clauses into the Water Act 1989 concerning the management and disposal of most of the 500,000 acres in the water authorities' ownership. The Forestry Commission – the largest public-sector landholder in Britain – was also targeted for privatisation. Since 1981 more than 2,600 woods – 46 per cent of the total once publicly owned – have been sold (see Table 8.2). Following a review of Forestry Commission incentives in 1994 the Government announced that the Forestry Commission would remain in the public sector – plans for wholesale privatisation of the Forestry Commission's land holdings were not pursued due to fears that the Government would be forced by public outcry to legislate for access to privatised woods. Yet despite this formal announcement the disposals programme has continued and sales have actually increased in pace.

There have also been attempts to privatise National Nature Reserves (NNRs) which were characterised by the media with headlines such as 'Batty Redwood to Privatise Snowdonia' (Lean 1995). Nicholas Ridley, whilst Secretary of State for the Environment, exerted pressure on the then Nature Conservancy Council to 'sell off' as many of the NNRs that it owned as possible (and declared a moratorium on the designation of new NNRs) (Ratcliffe 1989). John Redwood in his action plan for the Countryside Council for Wales requested that 'other organisations be sought to take over the management of NNRs' (Bishop 1997). Neither initiative had a profound impact as the respective Secretaries of State were removed from or left their offices shortly after launching the 'privatisation' initiative and their successors abandoned the initiatives.

The privatisation of public utilities, forced sales of local authority land deemed surplus to current requirements and continuation of the disposal programme for Forestry Commission woodlands coupled with reforms to agricultural policy aimed at making farming more market-oriented are leading to what various authors (Cloke and Goodwin 1992; Cloke 1992a; Clark *et al.* 1994; Curry 1994) have termed 'the commoditisation of rural space'. According to Cloke (1992a) this process of commoditisation has led to the production of an increasingly 'pay-as-you-enter countryside experience' with free informal forms of outdoor recreation replaced by a more formal, attraction-based experience which is paid for. Privatisation forces are but one part of this change in use and perception of the countryside, the move towards hedonistic forms of consumption being also part of this

Table 8.2 Summary of sales of Forestry Commission woodlands in Great Britain by country, 1981–96

	Total area 1981	*No. of woods 1981*	*Area sold 1981–96*	*Woods sold 1981–96*	*% of area sold*	*% of woods sold*
England	309,233	2,456	33,550	999	11	41
Scotland	728,753	1,704	75,501	820	10	48
Wales	149,725	1,718	15,867	861	11	50
Total	1,207,711	5,878	124,918	2,680	10	46

Source: Lean 1996

process. It can be argued that during the eighteen years of Conservative Government we moved towards a more exclusive countryside with use based on property rights or payment and notions of an inclusive countryside (e.g. the proposed legislation for commons and the Ramblers' Association's attempts to legislate for a 'right to roam') rejected.

Conservation policy: made in Britain, Brussels or Brazil?

The Thatcher and Major years were not marked by any great enthusiasm for the EU. The New Right vision of the Union, as developed in Mrs Thatcher's famous speech at Bruges, was a free market concept of a glorified customs union rather than a progression towards greater political, social and economic union. New environmental proposals should be subject to the closest scrutiny to ensure that an EU initiative was really necessary. Yet despite this 'Eurosceptic' stance, during the last 18 years, conservation of the nature and landscape of the British countryside has been increasingly influenced by factors that emanate from beyond the shores of Britain. The Common Agricultural Policy has long been acknowledged as an important influence on the shape of British agriculture, and through this on the countryside, but since the early 1980s membership of the EU (notably through the transposition of Directives) and wider international agreements and conventions (such as those agreed at the Earth Summit in 1992) have led to the establishment of new categories of protected area, the development of new strategies and action plans for conservation, increased regulation in the form of additional protection for certain areas and new procedures, and important changes in institutional arrangements.

Since 1979 five new categories of protected area have been, or are in the process of being, designated as part of the implementation of EU Directives and the UK Government has continued to designate certain areas as part of its commitment to wider international agreements (Biosphere Reserves and Ramsar Sites, for example). Nearly 330,000 ha of land has been designated as a Special Protection Area under the EU Directive on the Conservation of Wild Birds. The Convention on Wetlands of International Importance (more commonly known as Ramsar after the town in Iran where the convention was adopted) has led to the designation of over 80 sites covering nearly 330,000 ha. Although the number and spatial coverage of these European or international designations has increased considerably since 1979, the precise influence of these designations is confused by the way in which the UK Government has chosen to implement the various requirements in domestic legislation. Provision for the protection of all European and international sites is made through the UK designation of SSSIs or ASSIs (Northern Ireland) (i.e. a site must be designated as an SSSI or ASSI before it can be designated as an SPA or Ramsar Site, etc.). There has been no new primary legislation to implement the requirements of the Habitats Directive for example. However, implementation of the Habitats Directive led to the publication of a new Planning Policy Guidance Note on 'Nature Conservation' (DoE 1994c) and the introduction of new procedural measures into the planning system. These measures include charging the Secretary of State with responsibility for securing any necessary compensatory measures to ensure the overall coherence of Natura 2000 (the pan-European network of nature conservation sites of which Special Areas of Conservation will form a part) under Regulation 53 of the Habitats Regulations. This requirement, for the first time, formalises the concept of environmental compensation. PPG 9 (DoE 1994c) also outlines a very specific set of sequential tests for development proposals that might alter the status of a Special Area of Conservation. This is reflected in Annex C to the PPG which even includes a box stating that 'Planning Permission must not be granted'. The sequential tests and absolute prohibition of development represent a significant move away from the ideas of 'balancing conflicting material considerations' which have traditionally guided the determination of planning applications. The new guidance also introduces land management considerations into the planning process – a move that the Government had been staunchly resisting throughout the whole of the 1980s. Paragraph 23 states that 'Structure plans, local plans and unitary development plans . . . shall include policies encouraging the management of features of the landscape which are of major importance for wild flora and fauna' (DoE 1994c). The PPG also clarifies the requirement on ministers to confer with the European Commission before agreeing to harmful developments affecting European sites on grounds of imperative reasons of overriding public interest, where the Special Area of Conservation contains a priority habitat or species. Of all items of EU legislation to date, the Habitats Directive has had the most profound effect on the planning system. The provisions of PPG 9 represent fundamental changes to the principles that have underpinned the UK planning system: they significantly fetter the traditional discretion of land use planners and extend the role of development plans into land management.

The influence of EU membership extends beyond the boundaries of designations. EC Directive 85/337 on Environmental Assessment (EA) led to the introduction of new procedures in the UK. The Directive defines two types of development. All Annex/Schedule I developments (e.g. oil refineries, airports etc.) require an EA. However, Annex/Schedule II projects (e.g. afforestation, land drainage projects etc.) only need an EA if national or local Governments deem it to be necessary. Although, in principle, the EA Directive and subsequent UK regulations represent an extension of regulation, in practice the process has little impact on agricultural and forestry development. More significant from a countryside perspective is Directive 97/11 which amends the 1985 Directive. The new Directive requires that proposals for 'installations for the intensive rearing of poultry or pigs' are 'made subject to a requirement for development consent and an assessment with regard to their effects' and may require the UK Government to amend the GDO with respect to such developments.

Since the late 1980s, there has been a proliferation of conservation strategies and initiatives at the European and international level. In 1990, a European Ministerial Conference on the Environment adopted a European Conservation Strategy, which details broad conservation aims and is intended to provide a framework and stimulus for national policies. IUCN's Commission on National Parks and Protected Areas has published an Action Plan for Protected Areas in Europe. However, the most significant development from a UK perspective was the signing of the Convention on Biological Diversity at the Earth Summit in 1992. The UK Government played a constructive role in the Biodiversity Convention but there seemed to be a fundamental belief amongst key ministers and officials that this was a third world problem. Despite this 'post-imperialist paternalism' the then Secretary of State for the Environment, Michael Howard, was persuaded to convince the Prime Minister to declare that the UK would match the Rio outputs. This commitment initiated a formal process of biodiversity planning with the UK Government pledging to produce a biodiversity action plan by December 1993. The final action plan was published in January 1994 as part of a suite of post-Rio plans and strategies (H. M. Government 1994a, b, c and d). The 'UK Biodiversity Action Plan', despite its name, was not a plan, and this was recognised by the Government, following lobbying by key environmental groups (Butterfly Conservation *et al.* 1993). The final chapter of the 'plan' (H. M. Government 1994b) signalled the establishment of a Biodiversity Steering Group comprising academics and representatives from the statutory conservation agencies and non-government organisations to prepare a range of specific costed targets for key species and habitats for the years 2000 and 2010. The report of the Steering Group was published in 1995 and, although an official document, did not have policy weight because the Government was to prepare a response. The Government response was published in 1996 and marks a significant advancement in terms of conservation policy. Although not an Act of Parliament it was published as a command document and endorses the main proposals in the Steering Group report including the costed species and habitat plans. Aside from the obvious and potentially profound impact of the various individual action plans, this process has initiated a more formal system of biodiversity planning in the UK with a sequential, if non-statutory, model that includes the identification of measurable targets (a development proposed by the pressure groups in order to gain Government and Treasury support) with activity at national, regional and local levels.

Membership of the EU has also altered the institutional framework for conservation in the UK and often acted as a brake on the New Right. Implementation of the Birds and Habitats Directives has limited the discretion normally associated with the UK planning system and is shifting the balance of power from Whitehall to Brussels (albeit location- and circumstances-specific). The proposed development of Mostyn Docks on the Dee Estuary is a good example of this changed framework for nature conservation and its impact on New Right liberalisation tendencies. The proposed development of Mostyn Docks was within the Dee Estuary Special Protection Area and when John Redwood, then Secretary of State for Wales, declared that the application for expanding the facilities was only of local importance and would not be called-in he contravened the requirements of the Birds Directive. The Countryside Council for Wales had to inform the Secretary of State he was acting unlawfully and that development within such areas can only be allowed where there are 'imperative reasons of overriding public interest'.

The UK Government has often been slow or partial in its implementation of key EU Directives and international conventions in an attempt to curtail their influence on domestic policy and practice. For example, the UK has one of the lowest rates of SPA designation of all EU countries. Up to 1 April 1992, Special Protection Area (SPA) designation had been afforded to 37 complete sites and 10 part-sites – only 20 per cent of the total estimated internationally important bird areas (potential SPAs). Many potential SPAs were on estuaries and the Government may have been reluctant to designate such areas as SPA status could have threatened major infrastructure projects (barrages) and redevelopment proposals. In these and other circumstances, access to the European Commission and the European Court has been used by environmental interests as an appeal mechanism to challenge decisions made by the UK Government or lack of UK action. For example, the Royal Society for the Protection of Birds (RSPB) recently took legal action against the UK Government arguing that it had acted illegally when it left Lappel Bank out of the Medway Estuary SPA for economic reasons. The Advocate General's opinion and the European Court found in favour of the RSPB and the UK Government is now being urged to provide habitat compensation for the loss of Lappel Bank.

During the last decade control of large elements of environmental policy has passed from national capitals to the Commission and the EU (Baldock 1989). The Treaty on European Union signed at Maastricht in 1992 ushered in a quasi-federal system of policy-making through the introduction of Qualified Majority Voting for environmental policy. The UK

Government now faces the situation of having to implement environmental policies that its predecessor voted against in the Council of Ministers.

Conclusions

The central philosophy behind countryside policy has changed very little since the early years of the century. The leaders of Parliament's three biggest parties wrote to *The Times* on 9 February 1996 pledging their unity in the need to protect 'our countryside in its rich personality and character' – a pledge that repeats almost verbatim the hopes outlined by their predecessors in 1929. This consensus reflects the special place of the countryside in the British psyche. 'The countryside stands for all that is important in Britain; it is the expression of the good life away from the stresses and strains of the city and the symbol of everything which is considered truly British' (Best and Rogers 1973: 20).

Yet whilst the role and philosophy of countryside policy has remained remarkably consistent ideas about countryside conservation and the detailed mechanisms for delivering it have changed. Although the effects of the New Right 'liberalisation agenda' were not felt in the countryside to the extent they were in urban areas, it is possible to discern the development of a distinctive approach to countryside conservation based on New Right ideals of privatisation, market forces and limited state regulation. 'Enterprise conservation' (O'Riordan 1989) has been centred on a defence of the autonomy of private property rights and minimal state intervention. In contrast to urban areas where many of the planning reforms initiated by the New Right were concerned with weakening and constraining the powers of local planning authorities, in rural areas there has been a significant extension of regulations for countryside conservation. However, these new controls have normally taken the form of spatially or operationally specific controls and are often only available as a last resort. The prevailing view, as expressed by Nicholas Ridley (1992), has been that: 'the countryman in Britain knows more about preserving wildlife than the lot of them [environmental pressure groups, EC officials and conservation experts] put together' (p. 114). The emphasis on market forces has manifested itself in terms of new market mechanisms for the delivery of conservation and public access goals. Privatisation has taken place but by the back door through the Forestry Commission disposal programme and restrictions on local authority land holdings, for example. Conservation effort has been concentrated on the 'jewels in the crown' (primarily National Parks), creating what some have termed a 'differentiated countryside' where the countryside outside of protected areas has been more fully exposed to market forces. Despite the much-vaunted 'quangocide' of the early 1980s the number of non-departmental public bodies dealing with countryside conservation has nearly doubled. Enterprise conservation represents a *de minimis* approach to countryside conservation: policy developments have tended to be reactive – responding to problems only when the evidence becomes irrefutable – and the new forms of regulation that were introduced represented the minimum requirement. It is only in recent years that there has been a change of emphasis towards a more proactive form of policy-making (witness the Biodiversity Action Plan process and recent rural White Papers).

In tandem with the development of enterprise styles of conservation has been the growing significance of EU legislation which has often provided leverage against the Government and tempered the full force of free-market ideology. 'European conservation' has often been diametrically opposed to 'enterprise conservation': it is a force for fixed standards and processes rather than the administrative discretion associated with enterprise styles of conservation. The ideas and principles behind, and even some of the mechanisms for, countryside conservation are increasingly coming from Brussels or further afield. Transposition of the Birds and Habitats Directives has extended state regulation, required local planning authorities to formulate policies for the management of land rather than just its use, and established a supra-state system of site protection. During the last eighteen years control over large elements of environmental policy has passed from Whitehall to Brussels and a quasi-federal system of policy-making has emerged that contradicts New Right models of a free-market Europe operating as a customs union.

Although many authors point to Mrs Thatcher's supposed 'green conversion' in 1987 as a turning point for environmental policy, the influence of New Right thinking on conservation policy has been largely determined by the ideological stance of individual Secretaries of State for the Environment (and the territorial departments) and their ministers. There is also a difference in approach between the Thatcher and Major Governments with the latter years of 'Majorism' witnessing the development of a new approach to countryside conservation that attempts to consider the countryside as a whole.

9

THE GHOST OF THATCHERISM

Andy Thornley

Introduction

I would like to pursue two themes in this chapter. The first is to address the hypothesis of this book that the nature of the implementation process and the variation in local circumstances raise questions about the importance of Thatcherist ideology. To what extent does the research into the details of policy implementation show a divergence from the expectations raised by the ideology? Does this ideology fade into a blurred ghostly mirage when confronted with the facts about the nitty gritty actions of planners? The second theme, also raised by the editors in their introduction, is whether there were any significant changes when Mr Major took over. Mrs Thatcher's reign may have died but did the ghost of Thatcherism live on to haunt planning activity? In addressing these issues I will draw partly on some aspects of my previous work (e.g. Thornley 1993, 1996) and an analysis of the formulation of a planning agenda for London during the 1990s which I undertook with Peter Newman (Newman and Thornley 1997).

In this discussion it is important to distinguish between the level of setting the priorities and parameters of the policy framework and the details of policy implementation. The essence of my argument is that Thatcherism had, and in the 1990s continued to have, an extremely strong hold over the broader agenda-setting level. This means that discretion and variation in implementation may have been possible within this framework but the limits were tightly controlled. So if variation in the implementation is detected it is important to discuss the level of importance of the variable factors and how far they deviate from the overarching priorities. The strength and nature of the mechanisms of guidance, control and monitoring are key elements in the discussion. The basic priority of Thatcherism has been stated as the importance of the market as the decision-making arena and the procedures of Thatcherism as one of centralising power in order for this to happen. Now it may be said that these are very general characteristics and indeed it might be claimed that these are features of any government in the 1990s, even one led by Mr Blair. Putting aside the question of whether Thatcherism has had its influence on the Labour Party, the characteristics of Thatcherism achieve greater definition through examining the aspects which get a low priority such as decentralised democratic procedures, socially derived objectives and state intervention to safeguard certain longer-term public benefits – in contrast to Conservative governments in Britain since 1979 many modern governments are happy to espouse these objectives. Before discussing some of the features of Thatcherism which are most relevant to the above themes, a caveat is necessary. It is difficult to isolate ideology as an independent variable. It could be argued that ideology is merely a reflection of changing economic pressures, often now of a global nature. For planning there is also the question of the development market and how the changing demands of the market may cause particular responses in the planning regulatory mechanism. The Major governments have been operating in a different market climate from those of the 1980s. These economic forces are undoubtedly extremely important and any full understanding of the changes taking place in the planning system has to take them on board. Much can be learnt about the changes that have taken place in planning since 1979 through an analysis of ideological influence, though a fuller picture would need to also examine the interplay between ideology, economic change and interest representation. It can also be argued that limiting the study to Britain prevents an assessment of whether the development of the ideology is due to national conditions or part of broader international trends. Similar shifts towards a more free-market regime can be identified in many countries, most recently in Australia. It would be interesting to explore how far national circumstances such as political culture, history and political structures create variation in the way this free-market tendency is applied. These characteristics may also affect how deeply embedded the ideology becomes: for example, in Sweden a coalition of parties devoted to more liberal policies did not last long before there was a return of support for the Social Democrats (although this again raises the question of the relationship between left-wing parties and the New Right ideology). Thus a fuller picture would need not only to explore the interaction between international economic forces, political ideology and local power structures but also explore the influence of different national characteristics (for a study of these issues and their influence on planning across Europe see Newman and Thornley (1996)).

For the purposes of this discussion it is important to note that Thatcherism has in-built tensions. As a result contradictory approaches can develop. Sometimes efforts may be made to resolve the contradictions but at other times they may be left to co-exist. One possible comment on the difference between the Thatcher period and the 1990s is that whereas Mrs

Thatcher through her conviction politics was inclined to impose one line Mr Major did not (compare Mr Gummer's policy statement on sustainable transport in PPG 6 with other transport policies of the Ministry of Transport). This contradictory nature of Thatcherism has attracted considerable attention. It can be observed at the philosophical level (e.g. Levitas 1986) in which the neo-liberal and authoritarian strands are shown to have different views of human nature and the relationship between the individual and society. Promoting the freedom of the individual is at odds with the requirement for a strong authoritarian state. The tensions are kept under control through emphasising the common ground and the acceptance that a strong state is needed to provide the framework for market freedoms. However, the tensions are also evident at the level of interests. Historically the Conservative Party has always combined different interests – most notably the aristocratic landed gentry and the emerging capitalist class. At times the different factions that provide support for the party are seen to be in opposition. For example, during the 1980s those benefiting from the free market orientation, such as the housebuilders, came into conflict with those who wanted to preserve and conserve the status quo, protecting the environment and heritage from which they personally benefited. There has always been a conflict between the economic and environmental objectives of planning. As Fainstein and Fainstein (1982) have pointed out, the same urban space can be a requirement for capital accumulation while also being a territory in which people live. Thatcher governments were not immune from this conflict. They promoted both the ethic and the reality of owner occupation, which led to demands from the housebuilding industry for more development land, while also being subjected to pressure from wealthy residents who wanted to protect their pleasant environment. Some of the inherent conflicts in the planning process were particularly highlighted by the Thatcherist ideology.

Thus Thatcherism as an ideology is not stable and is subject to potential internal disruption. One way in which this was avoided during Mrs Thatcher's time was through her strong personal leadership, evoking outside threats and promoting nationalism. Maintaining cohesion was less successful under Mr Major. However, the strains in the Thatcher ideology were evident during her period of office and therefore the difficulties cannot all be blamed simply upon the change of leadership. Some of the internal ideological conflicts were becoming more problematic. This was evident in the pressures from outside to give greater consideration to the environment and the difficulty the government had in dealing with the factional conflicts between shire residents and housebuilders in the proposals for new settlement in the 1980s – as will be seen later, this problem led to the provision for greater local autonomy in local plan-making. The message emanating from these tensions in the ideology is that some variation might be expected as the tensions are resolved in different ways but this does not necessarily mean that the overriding principles of the ideology are sacrificed.

I would like to highlight another feature of the Thatcherist influence over planning. The basic framework of the planning system was kept in place, although severely eroded in certain parts. However, the nature of the British planning system allowed this framework to be moulded to a particular set of aims and priorities which emanated from the political ideological stance. The objectives and purposes of planning were changed even though the procedural framework remained largely intact. This was possible because of the high degree of control that can potentially be exerted by central government and the flexibility in the planning system which allows the same mechanisms to be used while the aims and objectives change. With the Thatcherist stance this potential was fully utilised, unencumbered by any constitutional constraints. The mechanism of planning policy guidance, appeals and other central government tools meant that the government could set its priorities and ensure that these were followed right down to the individual development decision. In this situation in which central government has such a high degree of control, through direct guidance, sanctions or last-resort intervention, decisions made at the level of implementation will be highly constrained. The Thatcherist ideology was not one in which discussion, local democracy, and listening to grass-roots or academic opinion were given any value. 'Bottom-up' influence on policy formulation was lacking. The rules were set and the criteria defined by central government. As I shall discuss later, it is possible to detect in later years a softening of this line with greater partnership in urban regeneration. However, for most of the Thatcherist period little critical analysis was commissioned to evaluate the success of government initiatives – even then the first such study was conducted by an arm of central government (Audit Commission 1989). This ideological environment conforms to a policy-making model which presents a top-down process of central formulation providing a given framework for local implementation. The criticism of this model which proposed a greater two-way process arose largely in the context of the 1970s. Since then we have experienced diminished local discretion and stronger central guidance. Centrally imposed legal, financial and organisation constraints have increased (Marsh and Rhodes 1992).

The ideological frame was set at a national level and clearly had a strong influence on national legislation, financial priorities and institutional procedures. This does not mean that there was no scope for interpretation within this national frame – particular power struggles between interests in a particular location or different local economic circumstances could create variation. However, the degree of variation was tightly limited by the ideologically influenced, centrally imposed, constraints. The strength of the constraints might have varied from one policy to another, or from one spatial area to another. One factor affecting the opportunities for local determination was the relationship of the particular policy to the potential tensions and contradictions that could occur within the ideology and whether these were reinforced by clashes between opposing interests. Thus different policy areas such as education, housing or planning may have experienced different degrees of central control and there may also have been variation within aspects of planning. Topics such as transport, conservation and retail location could be explored to test out how much guidance, control and monitoring was imposed from the centre. Similarly different regions and their local economic and political characteristics may also have generated differing degrees of autonomy, which might change over time. Thus Scotland in the early days of Thatcherism was less ideologically affected while in more recent times it was at the forefront of centrally imposed boundary changes.

Although, as I have already stressed, under Thatcherism central control increased and the opportunities for local policy determination diminished, this central control did contain variety. The ideologically derived principle of greater freedom for the private sector was not universally applied. Different strategies were adopted to meet varying conditions. As I have elaborated elsewhere (Thornley 1993), the Thatcherite planning approach can be described as the co-existence of three different systems. Most of the country was covered by a system in which planning controls were weakened and developers were given a freer hand to pursue their proposals. In this system applications still had to be considered through the development control process by the local authority. The powers available to the authority to intervene were eroded significantly. However, in less profitable areas the government recognised that even with this freedom the private sector would not instigate development. In these areas conditions had to be changed even further to encourage investors and developers to take the necessary risks. Central government then used its powers to change the conditions, such as financial arrangements, speed and certainty of decision-making, and the availability of cheap well-serviced land. Thus in areas covered by Enterprise, Zones, Simplified Planning Zones and Urban Development Corporations (UDCs) a different, even more slimline, planning system operated. However, there were other areas which were also treated differently. These were the areas where the demands for the protection of the built and natural environment were strong. The landed and owner-occupier interests in these areas were normally supporters of the Tory Party and were not happy with any relaxation of the planning system which would have a bearing on their amenity and property values. As mentioned above, the interplay of interests in these areas exposed a tension in the ideology. During the 1980s it was hoped that this tension could be held at bay by retaining strong controls in certain areas, creating a third kind of system. The incremental deregulation measures that were passed during this period excluded environmentally sensitive areas such as National Parks, Areas of Outstanding Natural Beauty and Conservation Areas. Thus the planning system operating in these areas still retained strong controls and over time became more distinct from the system operating elsewhere. However, as I describe later, this did not resolve the tension for long and conflicts erupted to cause considerable problems.

The rest of this chapter is divided into two parts. First I explore in more detail the question of whether there have been changes in the application of Thatcherism and particularly whether Mrs Thatcher's departure heralded a new era. Secondly I look at the changes that have taken place in the formulation of planning priorities in London. This will again illustrate how much change there has been over the last ten years. As it presents a case study of a particular area, it also takes the discussion down from the national level to a locality. It therefore provides material for the debate about local variation in the application of the ideology. Of course all localities have their particular characteristics, none more so than London, and so these local findings need to be placed alongside those of other areas to get a more complete picture.

The erosion of Thatcherism?

As is often the case with strong leaders, there was no natural successor to Mrs Thatcher and, unsurprisingly, the new Prime Minister, John Major, was a very different, some say rather grey and uncharismatic, figure. He was a consensus choice and can be said to reflect the view of some within the party that a period of calm and political peace was needed (Willetts 1992). So there can be no doubt that 1990 saw a sudden shift in the style of political leadership. However, it is necessary to separate the issue of style of government from its policy content. Does the shift in presentation mask a continuation of political purpose and strategy? Is Riddell right when he says there has been a change of 'personality and style, rather than of fundamental strategy' (1991: 220).

In exploring the ideological context for planning in the 1990s one has first to face the fact that under Mr Major there were no strong and clear messages of an ideological kind. In fact Mr Major, reflecting his personal political background, adopted the style of compromise within his party, although this was not an easy line to follow and his position was precarious. Such an approach did not lend itself to the propagation of a clear ideology. Thus on the level of ideological rhetoric there was a significant change. However, the question is to what extent did the Thatcherist ideological priorities still determine the policy approach of the 1990s notwithstanding this absence of rhetoric (the attempts of Mr Major to create ideological statements fell flat, e.g. the 'classless society' and 'back to basics' campaigns)? There have been suggestions that beneath this appearance of significant change, at the level of policy there was a considerable amount of continuity with the Thatcherist past (Kavanagh and Seldon 1994).

Within planning one aspect which would appear to contradict this thesis of continuity was the increasing concern over environmental issues. Did the environmental movement expose contradictions in the Thatcherist ideology? For most of her time in office Mrs Thatcher never showed much interest in the environment other than getting upset about the amount of litter on the streets. In line with the individualist approach, there was talk at one time of making people who had property fronting onto the street responsible for the tidiness of their patch. Because of Mrs Thatcher's conviction style of politics it is not surprising that the environmental movement had little effect. In addition the market orientation of Thatcherism did not lend itself to the financial commitments and longer-term perspectives required. The conservation of the environment implies some relaxation of the pure market approach. However, pressure to treat environmental issues more seriously was building up from two directions. The Common Market had been increasingly turning its attention to environmental matters. It had passed the Directive on Environmental Impact Assessment which the British government had adopted with some reluctance, and was exploring other environmental issues. For example, in 1990 it published its Green Paper on the Urban Environment in which it analysed the problems of cities and towns and put forward many ideas for priority action. Most of these required a framework of public intervention and planning as they could not be achieved by market processes. Their suggestions included increased urban densities, restrictions on the private car and better public transport, enhancing parks and open spaces, and protecting the visual quality and historic identity of cities.

A sudden change seemed to take place in the Thatcher government's approach to the environment in 1988 when Mrs Thatcher started to deliver speeches on the global environmental problem. However, these were directed at large world-scale issues and she advocated international approaches to deal with them. She was less interested in supporting policies of a national or local scale as this presented her with greater ideological conflicts. Meanwhile support was at hand in the work of David Pearce (Pearce *et al.* 1989) who had been asked to advise the government. He came up with policies to remedy environmental problems through taxation and modified cost–benefit analysis. This helped with the ideological difficulties as it suggested that market processes rather than public intervention could still be at the centre of decision-making. Subsequent Ministers therefore continued to stress that environmental issues could be solved by the private business sector, for example Michael Howard, Secretary of State for the Environment, said in 1992 that he wanted to put more emphasis on economic instruments and 'make the market work for the environment'. However, the European pressure did have some effect and in 1990 the government produced a policy paper (a White Paper) called *This Common Inheritance* setting out its environmental aims and including many of the ideas circulating in the EC (H.M. Government 1990). Although this policy has been monitored each year for progress, it has been much criticised for containing nice words but no financial support. The government continued with a low-key approach stressing litter, energy conservation and recycling.

Meanwhile a second source of pressure arose. During the 1980s the housebuilding companies had formed consortiums and used lobby pressure to try and get greater acceptance for the building of new houses in areas containing environmental restriction. This often took the form of promoting the concept of new self-contained villages with facilities. However, this generated a backlash of reaction in the areas where proposals were made – the well-known phenomenon of NIMBYism. As early as 1983 the government had tried to relax the constraints on developers through a modification of the Green Belt policy but this generated opposition from government MPs and party members who lived in the pleasant protected countryside. The same reaction occurred with the new village proposals and the government was presented with a split in its supporters between those who lived in protected areas and wanted the full range of controls to be retained and those who believed in greater freedom for enterprise. The problem was brought home to the government in the 1989 elections for the European Parliament when the Green Party gained an unprecedented vote of about 25 per cent in those areas threatened by new development. Many of these people would normally have voted for the Conservative Party.

Thus the combination of grass-roots reaction from their supporters with the demands from the EU forced the government into acknowledging the need to encompass the environmental issue and to present a greener face. This they did through a number of statements and policies, such as requiring Local Plans to include policies on sustainability. Local government also pushed central government along the environmental path, taking up the challenge of the Agenda 21 programme from the Rio Summit with enthusiasm. However, government adopted environmentalism without relaxing its commitment to the market as the prime decision-making arena and without devoting the necessary financial resources to back up the policies.

The 'new plan-led system' is another change that is usually mentioned to support the argument that the influence of Thatcherism on planning has ceased. The Planning and Compensation Act of 1991 gave renewed importance to the development plan and encouraged planners to enthuse about the start of a new era. The Act stated that planning decisions should be taken 'in accordance with the plan unless other material considerations indicate otherwise'. The plan is therefore the prime consideration, although the door is still open to uncertainty through the interpretation of the above phrase. The courts will be the final arbiter in the relationship between the plan and 'other material considerations' and it has been suggested that the plans will only have strength if they avoid being vague or ambiguous (Grant 1991). As mentioned above, the government has also extended the scope of Development Plans by requiring them to include a section on environmental sustainability. However, they have also reiterated the need for them to be 'efficient, effective and simple in conception and operation' and confined to land-use aspects (DoE 1992d) – thus reiterating the position taken in the early 1980s to tightly restrict the scope of the plans.

The move to a greater emphasis on local-level policy implied in the Act had been developing during Mrs Thatcher's last years. It was in 1989 that Chris Patten, the Secretary of State at the time, introduced the notion of 'local choice'. This arose as a strategy to extract himself from the problems over the new settlement appeals. At this time a number of applications had been submitted to build new villages in the countryside near major towns, particularly in the London area. The applications were usually opposed by local authorities and went to appeal where they then had to be decided by central government in a very exposed and publicised manner. Patten found himself caught between two lobbies, both natural supporters of the Conservative Party, the housebuilders and the residents of the shires. As already noted, the latter had been showing their displeasure through the Green Party protest vote in the European Parliamentary elections. He was in a no-win situation. The idea of 'local choice' allowed such difficult decisions to be shifted to the local level.

So development plans again became the arena in which the difficult job of balancing different interest groups could take place. However, how much autonomy would they have? Central government, having devolved this responsibility, still retained the ability to control and monitor the process. They could intervene in the preparation of plans if they thought the scope of the issues covered was inappropriate and they had the powers to call in the plan if it was considered controversial. Then, of course, they exerted much influence over the process through the Planning Policy Guidance Notes. Particularly important was the formulation by central government of strategic and regional guidance to which the Development Plans had to conform. Thus although Development Plans may have regained importance they could only use this power if they conformed to the boundaries set by central government, which had been strengthened in the last decade. Another conditioning factor on the plan resulted from the competition between cities. The desire of a particular city to promote itself in the game of attracting investment could condition the role and content of the plan. It may have become part of the city's marketing publicity, i.e. demonstrating that the city had a plan that encourages investment and provides locational opportunities that match the needs of companies and developers. Too strict a regulatory planning regime could divert interested parties to another city. Thus the local autonomy that had been awarded local authorities through the 'local choice' approach and the greater importance given to Development Plans has to be regarded as circumscribed. Freedom to formulate policies in the plan were highly constrained both by the boundaries set by national government, reflecting their ideology, and by the competitive economic environment which often determined local political priorities.

Significant changes also took place in the early 1990s in the approach to urban regeneration. During Mrs Thatcher's time policy towards urban regeneration followed a fairly consistent pattern. Central government would announce an initiative which they would administer and control through financial and regulatory powers. They would then use this power to open up decision-making to market influence and reduce local democracy. The Urban Development Corporations epitomised this approach. A major aim behind the initiatives was to provide the infrastructure, financial inducements and decision-making processes which would attract private-sector investment. Through creating the conditions which were attractive for the property industry it was expected that development would ensue and that this would create a spin-off effect. This property-led approach to urban renewal has attracted much critical attention (e.g. Turok 1992; Healey *et al.* 1992; Imrie and Thomas 1993), in particular for the way in which it ignores many of the dimensions necessary for city revitalisation and for its dependence on the cycle-prone property market.

The government has not put much emphasis on the monitoring and evaluation of its numerous initiatives. However, the two government-sponsored reviews undertaken both had critical comments to make. In 1989 the Audit Commission said that the 'programmes are seen as a patchwork quilt of complexity and idiosyncrasy' and called for a more coherent approach (Audit Commission 1989: 1). The government commissioned research to evaluate the success of the various urban policy initiatives implemented during the 1980s and this reported that the economic and environmental emphasis of the policies ignored problems of social disadvantage and that the local voluntary sector and local government should be more involved (Robson *et al.* 1994). The research could find no evidence that property-led developments had produced any trickle-down benefits for poorer areas.

In 1991 a new government initiative was launched called City Challenge which involved a number of new elements. One of these was the competitive bidding approach, since extended to other initiatives, in which local authorities were invited to enter a competition to try to win a limited number of awards. The approach has been criticised for involving a lot of time, money and effort in producing impressive bid documents and, as there have to be many losers, for often being unproductive. Government sets out guidelines for the bids and of course selects the winners, although they give no explanation for their choice. In setting out the guidelines the government made it clear that a high priority was placed upon attracting the commitment of private business to ensure good financial leverage and self-sustaining growth for the area. This requirement clearly influenced the choice of projects in the bids – those which would be economically viable rather than those which met the social needs of the area. The focus of the initiative on small areas also meant that problems which might pervade a wider geographical area could not be addressed. However, the initiative also put considerable emphasis on involving other agencies, e.g. the local voluntary sector, universities, Training and Enterprise Councils (TECs), and the local community. This greater involvement plus the local authorities' enhanced role in formulating the bid can be seen as a move away from the centrally directed approach of the UDCs with their divorce from local influences. However, in setting out the brief for the competitive game, central government is still able to impose its priorities. Many aspects of this new initiative show a shift away from the approach of the 1980s and have been welcomed by commentators, especially the incorporation of different sectors, the role of local authorities and the greater co-ordination between central government departments. However, many of the faults of the previous approach remain, such as the concentration on small areas and the limited finance. Considerable doubt has also been cast upon the degree to which the voluntary sector and local communities have been involved. Evidence suggests that, whatever the new rhetoric, the initiative was still geared to property-led physical regeneration. Although the City Challenge initiative did not involve any new expenditure, being 'top-sliced' from the urban regeneration budget, it fell foul of government public expenditure cuts in 1992. New rounds of the initiative were suspended. However, it was not long before the government announced further initiatives. These can be seen as the government's attempt to try and create better co-ordination between the different programmes, as suggested by the Audit Commission in 1989. There were two elements to this, a Single Regeneration Budget (SRB) and integrated regional offices of government departments. Both came into effect in 1994. The regional offices prepared an annual regeneration statement setting out key priorities, administered the SRB, and continued to be responsible for regional departmental programmes. The new budget, involving no extra resources, encompassed in one pot the myriad programmes of the five government departments involved – Environment, Trade and Industry, Employment, Education and the Home Office. The new budget continued the competitive philosophy of the City Challenge

Initiative; the government set out the guidelines and invited bids. Recipients were expected to make a significant contribution from their own budgets and maximise contributions from other sources such as the private sector and European funding.

There were a number of reasons for the new regional offices. One was to improve the integration of central government policy programmes. The better-integrated, regional administration also answered demands for greater regional devolution within England, and through the Integrated Regional Offices the government can match any *ad hoc* regional alliances set up by local governments. The deconcentration to regional offices did not, however, stem the debate about regional government, and it was a major issue in the Labour Party's 1997 election manifesto. The regional offices also represented a move towards a regional structure of administration better suited to the planning and funding regimes of the EU. Some local authorities feared that the regional offices would take a greater role in supervising EU funds.

Another initiative was also announced at the same time – it was called City Pride. Manchester, Birmingham and London were invited to compete for resources through preparing a 'city prospectus' in partnership with the business community, the voluntary sector, and public agencies such as the Training and Enterprise Councils. In the prospectus authorities were asked to set out promotional activities, a vision for the city over the next ten years, and a list of projects and how they were to be funded. There was an emphasis on economic objectives based upon a partnership with the private sector. At the regional level the Department of Trade and Industry introduced Regional Challenge in 1994. This further competition for development project funding was to be financed by 'top-slicing' European structural funds.

Did these newer initiatives indicate a change of direction? Certainly there was more devolution of responsibility to local authorities, who are responsible for formulating and co-ordinating the various bids. There was also a greater acceptance of the need to involve local communities and the voluntary sector. However, such local autonomy was again much constrained. Central government had a strong hold over the process through setting the guidelines and judging the bids and through tight control over expenditures through 'delivery plans'. It could also be argued that the government's priorities were still oriented towards creating the necessary climate for private-sector investment rather than addressing the social needs of the areas. If local authorities wished to win in the competitive game they had to show they were conforming to these priorities and local partnerships with the private sector helped to ensure this.

The case of London

A major change in the planning of London followed the abolition of the Greater London Council (GLC) in 1986. Strategic planning was fragmented into the thirty-two Boroughs and the City (the Corporation of London) who had to set out the strategic framework for their areas as the first part of the new Unitary Development Plans. The government also established the London Planning Advisory Committee (LPAC) to advise the government on matters that affected more than one Borough. This committee, comprising representation from all the Boroughs, has over the years produced its own statements of planning strategy for London; however, its ability to influence government has been constrained. Until May 1994, the political composition of the committee was balanced and therefore any statements had to attract cross-party support, thus limiting its scope. Even after the 1994 election in which the Labour Party gained control of many more Boroughs, doubts have been expressed about the ability of the London Boroughs themselves to take on a positive co-ordinating role across the city because of their diversity (Biggs and Travers 1994). The experience of other metropolitan areas suggests that if they are to do so a greater sense of metropolitan identity and more consensus would be required.

In addition many other bodies operated, not simply in an advisory role but to carry out London-wide services, creating a very complex picture (Hebbert and Travers 1988). Most of these bodies were appointed by central government and had little or no local democratic involvement. This was a national trend and its scale is illustrated by the analysis of Skelcher and Stewart (1993). They estimate that nationally, in 1993, there were 17,000 members of appointed bodies compared to 25,000 councillors, and they accounted for about 20 per cent of public expenditure compared to 25 per cent from local authorities. Skelcher and Stewart claim that at that time there were 272 appointed bodies in Greater London covering such essential services as education, health and transport. Although their figures may be somewhat overstated the point is well made that these undemocratic bodies cannot be considered exceptions but are central to an understanding of urban governance. As the same authors say, 'the effect of creating appointed bodies to provide local public services is to remove their policies and performance from the local political agenda. It raises the question of the nature of the democratic accountability of these bodies and their relationship to the Londoners whose lives they affect' (p.12). This shift to undemocratic agencies had an effect on planning as many of them covered policy areas with close links to planning activity, such as transport or economic development (e.g. London Transport and the Training and Enterprise Councils). The picture emerges of a planning approach scattered amongst numerous organisations many of which were undemocratic. There were 33 local authorities producing their overall plans, central government controlled bodies implementing urban regeneration, and appointed agencies producing and implementing sectoral policies.

On top of this institutional fragmentation lay a further layer of diversity. This resulted from the way in which finance was allocated for urban development. The concept of competition was now all-pervasive – as already mentioned there was not only the City Challenge approach which had been carried forward to the Single Regeneration Budget Challenge Fund but also Regional Challenge, City Pride and the Millennium Commission. This approach resulted in a lack of strategic thinking or co-ordination. There was no consideration of the overall effects for London of these different decisions and this contributed to a second layer of fragmentation. There was a further degree of randomness resulting from the investments emanating from the various National Lottery funds which often had planning and development implications. An indication of the resulting pattern of fragmentation in London can be obtained through a look at how some of these competitions and Lottery allocations affected London.

In the first round of the City Challenge two authorities in London – Lewisham and Tower Hamlets – gained funding and five more were successful in the second-round competition – Hackney, Newham, Lambeth, Brent, and Kensington and Chelsea. The transformation of this initiative into the Single Regeneration Budget led to a further fragmentation of the funding. Let us explore this through examination of the second round of this programme. The winners were announced on 12 December 1995. In London 201 outline bids were submitted, leading to 91 final bids of which 41 were successful. These schemes were to receive a total of £230 million over the following seven years. In announcing the results the Secretary of State for the Environment, John Gummer, said that the selection was a response to promoting London as a world city, reinforcing the Thames Gateway and tourist potential, while also responding to local needs. The six biggest schemes accounted for £133 million and the rest were scattered across London with about half the money going to outer Boroughs. The biggest allocation of £37 million went to the regeneration of King's Cross while £30 million went to refurbishing housing on the Roundshaw estate in Sutton. The other four large allocations were divided between schemes in the east and west of London, demonstrating the wide geographical spread. They covered the regeneration of Canning Town, infrastructure and training in the Hayes/West Drayton corridor and Wembley Park, and a cultural, tourist and environmental project in the Pool of London.

Funds available from the National Lottery extended further the fragmented and competitive nature of financial support. The fund was divided into several categories with different bodies involved in making the decisions – the Arts Council, the Sports Council, the National Heritage Memorial Fund, the Millennium Commission and the Charitable Fund. The whole process is another illustration of the way in which decisions were being taken in disparate, unaccountable, bodies. Although the Lottery funds did not have physical regeneration as a specific objective the investment of finance into winning schemes often had an impact on the economic and physical improvement of an area. The cumulative impact of the allocations needs to be assessed and related to other financial investments by government including those in the SRB or EU projects. The Lottery resulted in considerable funds being pumped into the London economy in an unco-ordinated way. Although most of the individual allocations were small and with little physical impact they could have an important cumulative effect on regeneration. There were also some bigger schemes with direct planning implications (e.g. £55 million for Covent Garden Opera House, £50 million for the Tate Gallery at Bankside, £12 million for the Shakespeare Globe). Even those schemes that failed in the competition may have developed local support for projects which went beyond the development priorities expressed in the relevant Unitary Development Plan. The Millennium Commission decisions in particular had a potential effect on planning and the Greenwich Peninsular Millennium project became one of the major development projects in the capital.

So a highly fragmented picture builds up with a large array of different *ad hoc* agencies, financial allocation through numerous different programmes and a plethora of discrete projects. However, at the same time there was pressure from many sides, including central government, LPAC and the business sector to develop an overall strategy or vision for London. The view was that London had to work harder at presenting its advantages if it was to survive in the increasingly competitive world in which cities try to attract the finite amount of inward investment. The promotion of London required a clear idea and framework for the future of the city. This need for a more cohesive approach led to a particular response in London, which was dominated by central government. Central government increased its involvement in the planning of London, leading to increased centralisation of decision-making.

One mechanism for this centralisation was institutional control. First, how did central government manage the activities of the London Boroughs? Local government in Britain was increasingly controlled from the centre from the 1970s. The Labour London Boroughs were participants in the major political conflicts with central government during the early 1980s and their ability to intervene in the urban environment was progressively weakened through controls on expenditure. Abolition of the GLC removed not only a tier of strategic thinking but substantial redistribution abilities. In the late 1980s overt conflict diminished and by the

1990s there were signs that local government had been let back into urban policy-making forums. The government encouraged the Boroughs to co-operate in a range of partnership bodies such as the Cross River Partnership, Thames Gateway and the Central London Partnership. The Boroughs were incorporated into the public–private London Pride Partnership in 1994. However, it will be seen that this new role for the London Boroughs was given on the government's terms. The content of a Boroughs' Unitary Development Plan was constrained by central government guidance – strategic guidance and the PPGs – reinforced by the appeal process which gives central government much power in the British planning system. The 1996 version of strategic guidance (Government Office for London 1996) was more detailed than previous versions. The guidance did not follow all of the advice given by LPAC (LPAC 1994; Hall 1995) and in particular stressed the importance of central London in maintaining and enhancing the world city role. The Boroughs, even through LPAC, played a secondary role to central government.

As already mentioned, central government increased its institutional control at the regional level to ensure that its objectives were implemented. In 1994 central government established a new Government Office for London (GOL). The new regional tier was established nationally but in London this tier fitted into a new political structure as a Minister was designated with special responsibility for London and a Cabinet Committee of senior Ministers established to co-ordinate policies in relation to the capital. It has already been noted that the reasons for these moves were many but included the desire to integrate ministerial responsibilities, to make it easier for government to pass down its priorities, and in the London case provide further symbolic proof of the government's commitment to the capital. The GOL had several roles in relation to planning. The first was issuing strategic guidance. The second was to promote the public–private London Pride Partnership within Whitehall. The third main function was managing regeneration budgets including the Single Regeneration Budget Challenge Fund.

How did these centralised controls deal with the issue of fragmentation? Urban regeneration initiatives were controlled by central government through allocation of finance and setting detailed rules and regulations. It might be said that from the City Challenge initiative onwards local authorities gained more power, as they could initiate the schemes and co-ordinate implementation. This initiative also included community representation. However, local authorities could only win in the competitive process if they followed the brief set out by central government which demanded a major involvement by the private sector and conformity with central government policy. City Challenge Initiatives and Training and Enterprise Councils had to be set up as companies limited by guarantee. It is said that this gave them more operational freedom. However, an alternative interpretation suggests that the company status led to greater secrecy and the interests of company finance came before any accountability to the wider community (Skelcher and Stewart 1993). Research has shown that the policy of involving the voluntary sector in these initiatives was not borne out in practice (Robson *et al.* 1994). The SRB projects had to conform to strict funding criteria and each successful project entered into a contractual relationship with central government to ensure the delivery of government objectives.

The 1990s witnessed an increase in the competition between cities as they struggled to attract companies in a more mobile economy. City marketing strategies arose as a response (Ashworth and Voogt 1990; Kearns and Philo 1993). As noted above, this was another pressure that led to the increased centralisation of decision-making in London. The city marketing approach was accompanied by changes in the governance of cities in which the development of urban growth coalitions was a common feature. Such coalitions in which the private sector played a leading role were first developed in the US and led to some well-known academic theorising (e.g. Molotch 1976; Logan and Molotch 1987). There has been a considerable debate in recent years about the relevance of this literature to the British and European context (Parkinson *et al.* 1992; Harding, 1994, 1995; DiGaetano and Klemanski 1993; Stoker and Mossberger 1994) and, in the development of regime theory, a broadening of the analysis to include a wider range of possible structures of governance (Stone 1989; Stone *et al.* 1991; Stoker 1995). The trends that underlie this analysis are evident in the London case. The central feature of Thatcherism in which increased centralisation is used in order to create opportunities for more private-sector involvement can be applied to this more recent trend of marketing the capital. Central government became increasingly involved in projecting the image of London and this contributed to the centralisation of the urban policy agenda. I will explore in a little more detail how the marketing approach in London has affected urban governance and increased the role of the private sector.

There was a growing consciousness on the part of central government, the business sector, the central Boroughs and LPAC, of London as a 'world city' with global competitors (Coopers & Lybrand, Deloitte 1991; London First 1992). For example, in the consultative pamphlet celebrating London's achievements issued in 1993 (DoE 1993d) the government expressed concern that other European cities were 'organising themselves to compete more effectively for inward investment' (p. 2). Ideas about a private-sector-led London-wide promotional body had been maturing in the minds of property and industrial interests (e.g. Robinson 1990; CBI 1991). Concern for London's world position was also shared by LPAC who co-sponsored the report 'London – World City Moving into the 21st century' (Coopers & Lybrand, Deloitte 1991). One of the suggestions in this report was the establishment of a promotional body, called the London Partnership, with the job of 'selling London's enterprise and culture, services and potential to the world at large' (p. 210). In their 1992 election manifesto the Conservative Party took up this idea and subsequently launched the London Forum to promote the capital as a tourist and cultural centre and to attract inward investment. Meanwhile the private sector set up its own promotional body, London First. In 1993 the two groups were merged under the banner of London First which is dominated by private-sector interests and funded by donations from companies. Thus the marketing of London is spearheaded by a private sector organisation with full central government backing and a good line of communication with Ministers.

When in 1994 the government invited the three cities of London, Birmingham and Manchester to participate in its City Pride scheme – involving a competition for funds through the production of a city 'prospectus' – the task of co-ordination in London was handed to London First. They brought in representation from the Boroughs and produced the London Pride Prospectus in January 1995. London First also set up an inward investment agency called the London First Centre. About a third of the funding for this centre came from central government in the form of the Invest in Britain Bureau. Although there had always been much informal contact between London First and the government, at the end of 1995 the government set up a more formal channel of communication with London First and the London Pride Partnership. They decided that a Joint London Advisory Panel should be established which would have regular advisory sessions with the Cabinet Committee for London. They also decided that this Advisory Panel would have the same membership as the London Pride Partnership, led by London First. The Secretary of State described this as 'the ultimate partnership' (DoE 1995d).

A further role adopted by London First, and subsequently the London Pride Partnership, was the encouragement of sub-regional partnerships which operated at a range of geographical levels. Some of the SRB projects covered fairly broad areas and generated inter-agency co-ordination. However, there was also a trend towards partnerships covering sub-regions of London. London First drew on the Business in the Community experience in other cities and conceived 'wedges' of London in which Business Leadership Teams could co-ordinate activity. In both east London and west London business alliances had preceded London First. The East London initiative grouped together local business leaders and the east London Boroughs of Hackney, Tower Hamlets and Newham in influencing investment policy and involvement in local business and community activities. The West London Leadership evolved around the Park Royal development and was extended to cover the corridor of London out to Heathrow Airport. London First promoted these sub-regional partnerships and encouraged similar arrangement in other parts of the capital (see Box).

Examples of London sub-regional partnerships set up during the 1990s

West London Leadership
East London Partnership
Central London Partnership
North London Leadership
South East London Partnership
South West London Partnership
Lee Valley Partnership
Thames Gateway London Partnership
Cross River Partnership
Deptford Creek
Wandle Valley
King's Cross Partnership
Gateway to Wembley Partnership

It can be argued that one of the reasons for the proliferation of sub-regional partnerships was to strengthen initiatives in the climate of competitive bidding and the desire to attract European funding. Many of the public partners believed that their 'real competitors are not so much other London Boroughs as the sub-regions of the European mainland who, on the whole have got their act together better than we have' (Stevenson 1994: 48). Whether or not other regions had 'got their acts together' the perception of business and public-sector leaders alike was that there was strong European competition. Central government was also fully supportive of this sub-regional development and the Government Office for London was involved in many initiatives.

Conclusions

Did Thatcherism continue to have a dominant influence over planning in the 1990s or did significant changes take place? First, there was certainly a shift towards a greater emphasis on the issue of sustainability. Environmental matters were never at the top of the agenda as far as the Thatcher governments were concerned and it has been argued that the adoption of sustainability was 'forced' upon government by international concern and the policy directives of the EU. Adopting and implementing environmental policies required taking a long-term view and challenging many vested interests. There is evidence that the government took the issue more seriously in the 1990s, for example there were Planning Policy Guidelines on out of town shopping centres and transport (DoE 1993b, 1994) which emphasised the need to adopt sustainable policies, the requirement that development plans cover such issues, and even government policy statements in favour of more support for cyclists. However, these initiatives were generally confined to statements of policy intent and did not explore in detail the implementation requirements and the financial resources needed. There was much public debate over what was meant by a 'sustainable London' but such issues did not figure prominently in the strategic documents and priorities of the government and supporting private-sector bodies. Thus in reaching a conclusion about the significance of the greening of Thatcherism it is necessary to explore further whether the commitment goes beyond a paper exercise.

A second dimension of change was the re-establishment of the Development Plan as a framework for decisions, which could be regarded as a reversal of the process of erosion under Thatcherism. Nevertheless it should be noted that the origins of this change were in Mrs Thatcher's period of office and cannot therefore be ascribed to a new climate of opinion brought about by her departure. This stronger role for the plan implied a change in the power relationship between developers and local authorities when decisions are taken over planning applications. However, it is important to remember the continued existence of central government strategic and regional guidance which had to be followed by local authorities in their Development Plans. This guidance which set out government policy for geographic areas originated in the Thatcher period and was the first time central government had itself produced a spatial policy – previously it relied upon reacting to plans produced by local authorities. The London case indicates that this guidance became more important – the 1996 Strategic Guidance for London was a much lengthier and more detailed document than previous versions. The private sector had an important influence over the content of this guidance. Thus although the Development Plan was a stronger instrument, this strength could only be employed if it conformed to the framework set by central government and its private-sector advisers. Another reason why the newly strengthened role of the Development Plan cannot be compared to the period before 1979 concerns the scope of the plan. In the early years of the first Thatcher government much attention was given to reducing the scope of the plan so that the social policies that many plans were seeking to include were removed and their scope clearly confined to 'land-use issues'. Although with the resurrection of the importance of plans in the 1990s their coverage was expanded to include sustainability and affordable housing, the government also reiterated the need to confine the plan's attention to land-use issues (DoE 1992g).

Urban regeneration policies indicate another area of change. The Thatcher period was dominated by the attitude that regeneration should be market-led. Property development was given a freer hand and the benefits were expected to 'trickle down' to the community. If the private sector was not keen on investing in certain areas because the risks were too high and the profit margins insufficient then the central government moved in and set up special initiatives to make the areas more acceptable to the private sector. Urban Development Corporations and Enterprise Zones were typical of this approach and a considerable amount of public money was spent in preparing the land and providing services. These initiatives by-passed local authorities and can be described as illustrative of the disdain for democracy within the Thatcherist ideology. A clear change can be seen in this aspect. The more recent regeneration initiatives of the City Challenge and Single Regeneration Budget were better integrated into the normal democratic processes. First, the local authority was involved. In the majority of cases it took the initiative in devising the proposal, co-ordinating the various actors and seeking funds from central government. The voluntary sector was also involved. There was therefore a much greater acceptance of democratic processes than in the 1980s and as a result a watering down of the authoritarian approach. However, again, certain caveats have to be applied. First, there was much criticism from the voluntary sector that their involvement was only token and their ability to influence the agenda very limited (e.g. NCVO 1993). The importance of local authorities in the exercise has also to be questioned. They were completely constrained by the way in which central government set out the brief for the competitions. This competitive approach, in which central government devised the rules and made the decisions without any need to provide reasons, gave central government controlling power. So, once again the local authority only regained responsibility if it accepted the priorities and agenda set by central government. The London case study reinforces this view that notwithstanding the devolution of some power to local authority to make greater use of Development Plans and to initiate urban regeneration projects, central government retained its controlling power. There was an ever-increasing fragmentation of institutions involved in planning and a proliferation of projects emanating from an increasing number of sources. This fragmentation was combined with an increase in central government control. Through its agencies such as the Cabinet Committee and the Government Office for London, its financial allocations, its rule-setting role in relation to competitions and its production of more comprehensive strategic guidance, central government increased its involvement. It strengthened its agenda-setting function, its determination of priorities and its monitoring of other bodies. It continued to give the private sector an important voice in the determination of priorities.

In considering the relationship between the political ideology of central government and the implementation of planning policy it is necessary to analyse the situation at different geographical levels. Each level could involve different characteristics of governance with differing degrees of central government involvement. There may also be variations in the interests involved and their relative strengths. I have also suggested that the approach of the government has shifted from one in which central government sought considerable control over the details of policy implementation to one of greater emphasis on formulating a framework which enables central government to control the agenda and determine priorities without necessarily being involved in the detailed implementation. Thus one would expect a lessening of central concern and an increase in local autonomy over the details of implementation but within a fixed set of parameters. This change in central government approach has accentuated the differences in the arrangements of government at each level. A distinction can be made between governance as direction-setting (defining objectives, identifying priorities and targets) and governance as steering (having day-to-day control of the rudder of governing organisations) (Rhodes 1996). Depending on the purpose of governance, the institutional arrangements and influences on decision-making will vary.

At the national level it is necessary to consider the legal framework which is established and national policy formulation. At this level throughout the 1980s there was a move to deregulation, orchestrated through increased centralisation (Thornley 1993). As we move into the 1990s we have seen that national government continued to be very active in formulating guidelines, controlling finance and ensuring that allocations conform to its ideological objectives. These objectives continue to give priority to market-based solutions and to give the private sector a strong voice in contributing to the policy agenda. One of the aims of this book is to explore whether an analysis of implementation at the local level leads to the conclusion that this broader framework is simply rhetoric and had little influence over detailed practice. Were actions taken at the local level able to run counter to the ideology? How much variation was possible between geographical areas? It can be argued that when we descend to the local level a wider range of agencies become involved in the implementation of policy and a certain degree of autonomy becomes available over detailed aspects within the centrally controlled framework. It is necessary to explore the way in which ideology is 'implanted' in particular localities and the potential for differential effects dependent on different local contexts and interests. To carry out such an analysis requires an exploration, amongst other things, of local politics and the particular power relationships. Regime theory, which builds on and expands the growth coalition approach, can make a useful contribution in such an analysis.

There is a strong correspondence between the emerging meanings of urban 'governance' in Britain and the focus in US urban analysis on 'regimes' (see Harding 1994). The commonality refers to the mixing of public and private power to deal with increasingly complex urban problems and the emergence of new networks aiming to provide urban leadership. Several authors have attempted to adapt the US regime literature for wider comparative purposes and a range of typologies of urban regimes has been produced to help understand changes in the governance of European cities (see, for example, Keating 1991; DiGaetano and Klemanski 1993; Stoker and Mossberger 1994). This literature raises several important questions in translating the US approach to European circumstances (Thornley and Newman 1996). These questions relate to the different relationship between central and local state, different financial arrangements, the links between business interests and political parties and leaders, the degree of local identification on the part of the business interests, and the strength of the public sector and professional and technical élites.

It is necessary to explore all these factors for a particular locality and explore the nature of the local system of governance or regime. However, the London example shows that, at least in a large city, it is necessary to consider different levels even within a locality. Urban regime theory has been developed and based upon an analysis of medium-sized cities in the US. A more complex picture emerges when analysing a city the size of London. Utilising the approach of regime theory it is possible to suggest that different kinds of regime could operate at different levels. The national level which sets out the legal framework, the national policy guidance, the funding priorities and the brief for urban initiatives and allocation of the funds clearly affect London as much as anywhere else. At this level the decisions are obviously taken by central government. Moving down to the London-wide level, the national framework is adapted to London circumstances and it is possible to identify an urban regime which is in the business of formulating priorities, agenda-setting, devising visions and promotion, for the geographical area of the capital. The dominant interests in this regime are central government strengthened through its regional arm, the Government Office for London, and the private sector that is given a leading role in the form of London First. Other interests, including local government, are drawn into the process but in a more advisory capacity. Although dominated by central government, the attention given to city marketing and promotion resembles a variation on the US 'growth coalition' model. At the subregional level the balance changes with less central government involvement and a greater emphasis on the partnership between local authorities, other bodies such as TECs, and the business sector. However, these subregional regimes have to operate within the strategic guidance, legal and funding rules set by higher levels. Again it might be said that this level could also be described as a growth coalition – this time of a more traditional form as central government is less involved. However, the particular orientation of the coalition is towards funding opportunities rather than towards attracting company investment. At the level of the Boroughs a different regime can be detected: here the local authorities play a dominant role with strong links to community interests. In contrast to the London-wide and sub-regional levels, this level has a direct democratic input. This level has the responsibility of formulating the newly strengthened Development Plans. However, as they operate within a given strategic framework their autonomy is limited. Many areas where more extensive change and public investment are occurring are being implemented through the special arrangement of the urban regeneration initiatives such as City Challenge companies and Single Regeneration Budget teams in which a wider range of interests are involved. These project-oriented arrangements could be described as another kind of regime – a development or implementation coalition – in which the partners in the coalition take on yet another arrangement with central government again playing a major role. Once the development areas associated with special urban initiatives are removed from their direct authority, Boroughs themselves could be said to be mainly concerned with conservation and detailed development control, interacting with heritage associations, residents groups and NIMBY organisations. Such a regime has been described as a caretaker regime (Stone *et al.* 1991).

Thus in London it is possible to identify different kinds of regime operating at different levels and in each case the influences on planning and the role it undertakes differ. For example, at the strategic level planning is involved in the process of defining priorities and contributing to the evolution of a vision for the future. This level is dominated by central government with the advice of the private sector and is geared to international city competition and the desire to preserve London's world city status. Sub-regional strategies are devised for areas that cover several Boroughs and involve strong inputs from quangos and the private sector as well as the local authorities. Development Plans are formulated at the level of the Boroughs themselves. Here the local authority has more influence but within the framework set by central government guidance. It could be said that the main characteristic of these plans is the preservation of the status quo. Areas with potential for change are likely to be covered by special initiatives. Funding requirements lead to a strong central government role in these areas and the Boroughs monitor the planning process closely insisting upon a strong private-sector involvement.

Local implementation was highly constrained by central government ideology from 1979 to 1996 although the mechanisms for achieving this changed during the period. Central control of local policy became less direct and more subtle but remained a dominant influence nonetheless. The ideology was less stridently pursued but the importance of the market mechanism and the centralisation of decision-making remained at the heart of the Major governments' programme. The move from an ideologically overt and involved approach to one involving a less outspoken rhetoric and more strategic control still only allowed a limited degree of local autonomy. However, increasing difficulties in containing inherent conflicts sometimes provided opportunities for local initiative. Nevertheless the principles and priorities of Thatcherism continued to inform the government's approach in the later years although they were presented with less consistency and confidence. The governments of the early 1990s built a different kind of policy edifice concentrating on strategic intervention and monitoring. However, the ghost of Thatcherism still haunted the structure, ensuring the continuation of the ideological agenda. It will be interesting to see whether the new occupants of the building can completely banish the spectre of Thatcherism.

Acknowledgement

I would like to thank Peter Newman for his valuable contribution. The material for the London case study is drawn from our joint research on planning in London (for further details see Newman and Thornley 1997).

10

CONCLUSIONS

Philip Allmendinger and Huw Thomas

Introduction

Even before this book was published it was clear that the idea of the 1980s' witnessing the death of planning was wide of the mark. The rhetoric of the time was confrontational, the ideology and ideas appeared antagonistic and some of the actions seemed to herald a new kind of non-plan. But planning survived. Since 1990 we have had the 'plan-led' system as well as commitments to concepts such as sustainability which rightly place land-use planning centre-stage. However, even this 'renaissance' has been a double-edged sword (of which more later). So, two prime ministers on since Mrs Thatcher first came to power in 1979, are the intentions and actions of the New Right no more than a historically interesting but insignificant story? We feel not. This book is not simply about the New Right approach to planning and why that approach turned out differently than expected (though that in itself would justify any interest). The wider picture that the contributory chapters have aimed to portray is the nature of planning practice in the UK in its political and administrative diversity. The New Right provides a useful though not crucial perspective. What made its programme particularly illuminating was the stridency of approach coupled with the eschewing of compromise and consensus. However, a similar though less conclusive picture could have been expected of other governments in similar positions: the Tories merely shone a particularly bright light on planning as well as on themselves. So, we can learn as much about planning and its context as about the New Right and its approach to policy and implementation. This is particularly important given the recent change of government – with no experience of power since 1979 the Labour Party need to recognise limits to change. What we aim to do in this chapter therefore is to examine the New Right from a multi-faceted perspective:

1 What were the impact and influence of New Right approaches to planning locally (and by implication nationally)?

2 How do ideas of implementation and localities help us understand the above?
3 What does this tell us about the political, administrative, spatial and autonomous nature of planning practice in the UK?

All of the above can be summed up in one word: change. Where did the ideas for change came from, what happened to the changes introduced, and how did the planning and UK local government mechanisms mediate and resist that change? Before we address how much change it will be worth reminding ourselves of the radical nature of the New Right challenge.

Although Marsh and Rhodes (1992) come to the conclusion that the New Right is a diffuse phenomenon with contradictory constituents, virtually all would agree that the policies and politics that emerged after 1979 constituted a distinctive phase in the history of post-war British politics (Savage and Robbins 1990: 1). King (1987), Marsh and Rhodes (1992), Kavanagh (1987) and Gamble (1988) among others are certain that the government entered office in 1979 with a clear commitment to the liberal and authoritarian strands of Thatcherism and that there is little doubt that these influenced the policy agenda. For example, the government's economic policy reflected the liberal strand and was based on monetarist principles designed to reduce inflation at the expense of higher unemployment. The subsequent switch to privatisation (Jackson 1992; Johnson 1991) aimed to reduce the public sector and public spending and to deregulate the market. Trade union legislation was based on a combination of liberalism and authoritarianism and was an attempt to reduce the unions' influence on the operation of the market, regain control and reassert the government's authority (Marsh 1991). A number of studies have sought to measure the extent of policy change in the New Right era and link this to their aims and philosophy (for general accounts see Cloke 1993; Kavanagh and Seldon 1989; Marsh and Rhodes 1992). All of the studies agree that a great deal of legislation was introduced during the period, much of it very radical (Marsh and Rhodes 1992: 170). However, many studies of the New Right *policy outcome* have questioned how much was achieved: 'The Thatcher governments may have had more radical objectives than previous governments, but they were probably no better at achieving those objectives' (Marsh and Rhodes 1992: 170).

Burton and Drewry (1990) conclude that more legislation was introduced by the Thatcher governments than in previous comparable periods. Although these changes were enacted without difficulty the government failed to achieve many of the aims it set itself. Obviously it is difficult within the scope of this work to review all the New Right's policy areas, although others have done this admirably (Kavanagh and Seldon 1989; Cloke 1992; Marsh and Rhodes 1992). Nevertheless, even in those areas that the government regarded as most important their achievements have been much less than claimed.

Studies by Bradshaw (1992) and Wistow (1992) illustrate that in relation to social security and the National Health Service the government took an electorally popular rather than ideologically driven line. Marsh (1991) is in no doubt that the failure to introduce more competition into the government's privatisation programme of nationalised industries was due to a belief that greater competition would reduce its attractiveness and the financial return to small investors and thus be electorally damaging. Bradshaw (1992) has demonstrated that the philosophy of the government was less important in initiating legislation in the field of social security than demographic changes, while Wistow (1992) concludes that a combination of demographic trends and medical advances exerted an upward pressure on medical expenditure. Marsh (1991) has also demonstrated that the change in policy from monetarism to privatisation came about because of the failure of monetarism to reduce inflation rather than for philosophical reasons.

The image of Thatcherism in particular as embodying conviction politics as described above is also brought into question by McCormick (1991) and Ward and Samways (1992). Environmental pressure groups exerted enough influence to 'green' the Tories while negotiations with the unions led to clauses of the 1984 Trade Union Act being removed (Marsh 1992) and managers of recently privatised companies successfully lobbied to retain their monopolies (Marsh 1991). Marsh and Rhodes (1992) also point to two other influences on government policy. The first was the European Community which increasingly influenced environmental, agriculture and economic policy areas and the second was the role of Parliamentary Committees which, as Wistow (1992) shows in the case of the Social Services Select Committee, directly changed policy.

As a result of such studies Marsh and Rhodes (1992) come to the conclusion that too much emphasis has been placed on the ideological and theoretical aspects of the New Right at the expense of studies concerning the extent and effect of policy change. The question therefore is: to what extent has this been reflected in planning?

How much change?

It is clear from the introductory chapter and the discussion above that different perspectives have always existed on the extent of change. However, the chapters in this volume point to a consistent under-achievement during the Thatcher and Major years. In the field of housing, which was one of the most contested and sensitive areas of planning, Bramley and Lambert paint a picture that seriously calls into question notions of radical difference. Some major ideologically driven changes such as a reduced role for the state in providing and financing housing were ruthlessly effective, though attempts to deregulate and reduce the bureaucratic burden met with more mixed success. Financially led changes such as reductions in subsidy to public housing were allied with the popular and successful 'right to buy' scheme. But, as Bramley and Lambert point out, market mechanisms have always dominated the land and development process with planning actually supporting the market. Further (and echoing other analyses of the period), those who benefit from planning controls have traditionally been more middle-class, (C)conservative-minded voters. Thus the planning system was to be used to facilitate rather than hinder the extension of public ownership. Some of the major changes to this end included housing land studies involving housebuilders with the requirement that local planning authorities ensure a five-year supply of land and that market criteria be used in determining availability. These changes were balanced by the need to ensure that environmental considerations were taken into account. Other measures of deregulation similarly had mixed success, e.g. the White Paper on the Future of Development Plans.

Hull and Vigar highlight the changes proposed and introduced by the Conservatives with regard to development planning. The first category of change included alterations in primary legislation to either speed up or by-pass aspects of plan-making. Second, changes in representation or interpretation through guidance again pushed for speedier or less cumbersome planning. Finally, the government used its executive position to enforce these and other changes. Contradictions in this approach have appeared that can be traced back to liberal and authoritarian tenets of Thatcherism but these were compounded by central and local-level conflicts and the vagueness of policy guidance. Far from making the planning system more responsive the discretion still available to some local planning authorities made sites more difficult to develop. To a certain extent the effectiveness of the New Right's approach depended on the locality as well as the particular institutional and regulative framework in operation. So whereas environmental concerns were of particular relevance in Kent and the plan-making process acted as an effective conduit for this the same concerns were both less visible in the West Midlands and less likely to emerge there because of the unitary plan framework and the role of the Government Office for the West Midlands. Of particular interest, however, are the changes introduced in 1991, i.e. Section 54A or the 'plan-led system'. On the surface the approach appears to herald a pro-planning stance though evidence now suggests that this was simply another way of streamlining the development plan process. If this were the case then clearly it has backfired. Hull and Vigar back up anecdotal evidence which demonstrates the vastly increased time taken to adopt a plan coupled with the delay and resistance to pursuing major projects in the absence of any plan. This has recently led the government to make sweeping changes to plan adoption procedures (see *Planning* 1214). Part of the reason this has emerged as a problem and thwarted attempts to speed up, simplify and reorient the process is the local planning authority's ability of self-adoption and the non-binding nature of inspectors' recommendations.

Conservation has always been a contradiction for the New Right – ideologically strong in the Conservative tenet and politically popular throughout the 1980s and 1990s. It is conservation in its widest sense that has provided much of the tension with free-market liberalism. There is little doubt that conservation and heritage are deeply embedded in the nation's psyche and are popular (particularly in the Conservative-minded shire counties). Such widely supported feelings are shared with a large proportion of the Conservative Party and have led to damaging criticism of any proposals perceived as in any way threatening. This has not stopped the New Right from attempting to deregulate, relax and by-pass conservation controls. But, as Larkham and Barrett point out, tension within government and particularly between different government departments, as well as between central and local governments, has diluted or thwarted proposed changes, notwithstanding some individual, almost perverse, decisions on listed buildings. Regardless of the individual peculiarities of characters such as Nicholas Ridley, Jocelyn Stevens and Teresa Gorman it is difficult to come to any conclusion other than that conservation and heritage concerns have remained virtually unscathed throughout the New Right years – and this despite there being few new positive powers for local authorities, decreases in funding and a general centralisation of control.

Philip Allmendinger charts what he terms the high and low points of the New Right approach, Simplified Planning Zones (SPZs). Potentially a radical departure from the traditional separation of plan and permission characteristic of the UK system, SPZs provide another clear example of thwarted and diluted intentions. But, much more than this, they also demonstrate a spectacular misreading of the market-supportive role of planning. The idea was that if property interests were given the opportunity to simplify the planning system they would jump at the chance (though, as the chapter demonstrated, the mechanisms were in fact far from simple). This was shown to be very wide of the mark. As many private-sector interests concluded, they preferred the erstwhile discretionary regime and the certainty it involved. As a consequence of this, and the complexity of adoption procedures, only half a dozen or so zones were ever adopted. Like design and development plans, the vagueness of government guidance allowed different and often contradictory interpretations.

Elizabeth Wilson maps out what has been the Achilles heel of New Right thinking – the environment. If any one topic has been a running sore for the New Right it is the environment. But, as Wilson concludes, the source of this is not only the contradictions within the New Right between the free market liberals and the conservationary-minded Conservatives, it has also been between different interpretations of what liberalism is, the nature of UK planning and the growing influence (a sore itself) of Brussels. Tremendous increases in awareness and policy activity cannot be separated from the wider public and intellectual concerns – the New Right after all had no divine right to power and still (especially after 1987) needed to win elections and hold together support.

Whatever qualifications there may be when judging the extent of change to other features of the planning system there can be few doubts that development control changed under both Thatcher and Major administrations. 'League tables' based on the speed of determining planning applications, planning fees, accusations of jobs 'locked in filing cabinets' and concern about achieving 'quality' (however defined) are just some highlights of the ferment of debate over development control – perhaps the most visible aspect of the planning system – in the 1980s and 1990s. Tewdwr-Jones and Harris argue that the mechanisms through which central government has tried to achieve change have changed in the 1990s, as legislative and policy prescriptions are supplemented by more generally defined initiatives such as the introduction of the Citizen's Charter into planning. But one thing remains constant: pressure on local planning authorities to undertake development control as central government wants it done. However, Tewdwr-Jones and Harris also make it clear that planning authorities have not responded uniformly to these pressures, and they point out the significance of local political circumstances in shaping how individual authorities have reacted to the intense central government interest in development control.

On the whole, the impacts of the New Right on planning are broadly consistent with their impacts on most other policy areas – there has been some change but not as radical as the rhetoric or dogma would have indicated. But there remains the question: why? Some of the contributions point to contradictions and confusion in the ideology and approach of the New Right. The fusion of liberal and conservative ideas made the New Right radical but also created tensions and problems. What united them (the common enemy of socialism and the need for change) was not enough to overcome differences in detailed policy prescriptions. In some policy areas either a market solution or centralisation was appropriate; e.g. in privatisation of the utilities or in increasing public spending. Conflicts here were minimal. In some areas a fusion was possible – trade union legislation which pleased free-marketeers and strong-state Conservatives. But in numerous policy areas where overlap existed no such happy compromise could be reached. In planning one could point to environmental and conservation concerns as prime examples of different solutions to the same problem from within government. One of the consequences for a radical government intent on some change was the need to phrase policy and legislation in such a way as to satisfy both camps. The result was vague and ambiguous policy objectives and guidance: the need to create 'certainty and flexibility' through SPZs – an oxymoron that was never resolved.

Another reason why much less was achieved by the New Right than they might have expected was highlighted by the attitude towards the planner's role in relation to the market. Much of the rhetoric aimed at planning, particularly in the early years, concerned its interference with the market. The answer was to deregulate and release pent-up demand for investment through a 'bonfire of controls'. As Bramley and Lambert point out, this failed to recognise how planning actually supports the market through creating certainty for investors. As many property owners and developers recognised, limiting supply through planning allows investors to gauge with greater certainty the profitability of a scheme. It also allows reaction to the plans of others through consultation on plans and applications. This is one of the main reasons why there was not a clamour for greater deregulation of planning from private sector interests, only for greater speed and certainty.

Although the market-supportive role of planning blunted the liberal critique and approach this was bolstered by a much more sympathetic and supportive perspective from the authoritarian strand. We must also not forget that not all of the government's supporters throughout the 1980s and 1990s were of what could be identified as the New Right. Many were of the older 'one-nation' Tories known by Mrs Thatcher as the 'wets'. Although not popular with the government these MPs made up their majority and needed to be 'on-board' in policy as much as either the liberal or authoritarian factions.

These then provide some reasons why the New Right achieved less than they would have wished. But the Introduction highlighted another less obvious reason: the New Right's approach to implementation.

Implementation and change

The Introduction identified the possibility of an implementation perspective providing a useful insight into the extent of change under the New Right. The prescriptive top-down perspective characterised by Hogwood and Gunn (1984) and Sabatier and Mazmanian (1979) fitted in well with the strident conviction policies of Mrs Thatcher in particular. Centralised policy formulation, control and implementation characterised the New Right governments:

> In effect, the Government operated a top-down process model of policy making in which it could, and should: set the policy agenda and choose the policy options, unencumbered by the

> constraints provided by interest groups; pass the legislation without amendment, given its majority in Parliament; and control the implementation process to ensure that its objectives were obtained.
>
> (Marsh and Rhodes 1992: 8)

The evidence from the chapters in this volume is that the New Right's choice of a top-down approach to implementation led to what Marsh and Rhodes (1992) would term a 'self-inflicted implementation gap' (p. 9). The six constraints of the top-down model identified by Sabatier (1986) provide a useful starting point:

1 clear and consistent objectives
2 adequate causal theory about the situation and its causes
3 appropriate policy tools and resources
4 control over implementing officials
5 support of interest groups and agencies affected by the policy
6 stable socio-economic context which does not undermine political support.

Taking each one of these in turn we can see how the government failed in all six.

Clear and consistent objectives

As mentioned above and by the various authors throughout this volume, one of the main impediments to clear and consistent objectives was the ideological and very real contradictions within the New Right. However, it is easy to see this as too clear a picture and, as Bramley and Lambert point out, all governments involve a degree of pragmatism aimed at electoral success. Nevertheless, it is clear that different policy prescriptions do emerge. Gamble (1988) may have termed it 'free economy, strong state', but the liberal perspective would be 'free economy, minimal state', while the authoritarians would go for 'mixed economy, strong state'. Of the six main elements that would have constituted a theoretical New Right approach Bramley and Lambert found that only one – the reduction in the direct role of the state in housing provision – has been unambiguously and consistently implemented. One of the reasons they identify for the failure of implementation is the lack of clear objectives. Allmendinger sees the main difficulty here being in steering a course between the two tenets to encompass both and alienate neither; the result is bland general objectives. Government guidance will inevitably be vague to a degree but this is normally because of the difficulty in being prescriptive about disparate situations and conditions. But the vagueness emanating from the New Right compounded this in the lack of clear direction for policy. Bishop notes the implications of internal tensions within the New Right in relation to countryside conservation – while some factions wanted a relaxation on planning controls in the Green Belt, others resisted. Added to this we must recognise the variability of external and internal influences (covered more fully in the final subsection of this section). This is highlighted by Hull and Vigar who point to the 'U' turn in government attitudes towards development plans.

It would be misleading to talk of 'U' turns in relation to development control, but Tewdwr-Jones and Harris point out the tension between the emphasis on speed, efficiency and providing a service to clients so characteristic of the early and mid-1980s and the exhortations of the late 1980s and the 1990s to improve the quality of the service (and the built environment). These tensions remained unresolved, not least because of what was meant by 'quality' and how it might apply to the various parts of the planning process.

Adequate causal theory about the situation and its causes

This area of implementation failure is more difficult to ascribe to the New Right due to the normative nature of their causal theory (or theories). The change in emphasis in monetarist policy in the early 1980s from using £M3 to £M1 and eventually £M0 as a measure of money supply reflects an evolution in macro-economic policy rather than a coherent strategy (Jackson 1992). But some of the assumptions behind changes to planning demonstrated a lack of understanding of the role of planning in markets. A blinkered belief in the supremacy of unfettered markets led to a simple approach: deregulate as far as possible and wherever possible. Similarly, where state intervention was necessary it should be more centralised. The problem, as we have already touched upon, relates to the market-supportive role of planning. The theory that markets are best left alone does not account for the symbiotic and evolutionary nature of the relationship between markets and intervention in the form of planning. The New Right portrayed markets and intervention as separate – you could increase or decrease either with a corresponding effect on the other. After fifty years of land-use planning the relationship is now more like an alloy of metals – two separate constituents but fused into a different object. Too many investment decisions, such as house prices, are now heavily dependent upon there being a planning system. It is no longer possible to separate the two as they really do not exist as separate entities any more. Ironically, it was the centralisation aspect of New Right thinking that was most successful. However, this could not be regarded as a causal theory but simply as an ideological or normative belief.

Appropriate policy tools and resources

The New Right inherited a planning system that had a great deal of flexibility built into it. As a product of the post-war social-democratic consensus and the UK administrative tradition it had a distinct balance between the centre and the local which favoured the former. The New Right were therefore able to make changes without primary legislation, but such changes were made through national policy guidance. This guidance was (as we have discussed above) broad by its very nature. It was therefore difficult to control from Whitehall what went on locally. The tacit assumption (particularly in the early Thatcher governments) was that central policy change would equal local policy change. The evidence from the chapters points to a different picture. Attempts to downgrade the importance of design through policy guidance ignored its inherently local and subjective nature. It also failed to recognise that attempts to reduce the power of local planning would not be achieved by reducing 'material considerations' such as design. The powers of planning locally lie as much in procedures as policy. The only way to enforce policy changes is through the appeal process – something most developers and applicants will avoid if they can. But the threat of appeal and long-drawn-out and expensive inquiries is enough to give power to local planners to still include matters such as design. A lot of developers will be happy to negotiate on design in the knowledge that planners do not have policy backing, if only to avoid an appeal situation.

The same complex response (and incomplete adherence to) changes to guidance can be found in other policy areas: housing, businesses, speed of decision-making, consultation, etc. However, the New Right did achieve greater success in implementing their ideas (and those of the EC) when changes were made to statutory instruments. For example, Environmental Assessments are now a common feature of the planning system. Whether this is because the use of secondary legislation is more effective than policy guidance or whether planners are more sympathetic to the aims of Environmental Assessment is unclear, though it is likely to be a mixture of both. This merely points to the poverty of a purely 'top-down' perspective that excludes or alienates those actually charged with implementation. In the case of primary legislation, again a mixed picture emerges. Yes, initiatives such as Urban Development Corporations, Simplified Planning Zones and Section 54A were introduced through primary legislation but the 'success' or evolution of these initiatives was very dependent upon other matters such as the accompanying policy guidance. Thus the New Right attempted on the whole to change the planning system with existing mechanisms, often misunderstanding the nature of the issues and where power, policy interpretation and influence lay.

Control over implementing officials

The literature on street-level bureaucrats is particularly relevant to the role of planners in relation to the New Right approach. Lipsky (1978) has persuasively argued that the role of street-level bureaucrats in carving out autonomy and decision-making power in public bureaucracies effectively sets them apart from the organisation itself and makes both reform and implementation difficult if they are not 'brought along' with the proposals. This point is made particularly strongly in the case of Simplified Planning Zones in this volume where discretion, information control and decision-making power combined with vague policy guidelines allowed planners to radically alter the shape, content and direction of policy.

In the case of development plans Hull and Vigar specifically examine the role of local discretion as compared with centralisation. The ability of planners to tell two totally different stories to developers and conservationists about the same site clearly demonstrates the scope for discretion and power play. Similarly, the vastly different approaches to strategy formulation in Kent and Lancashire highlight the ontological and inevitable variation in planning practice notwithstanding the centralisation of control that has occurred during the past eighteen years or so. Associated with this is a high degree of illusion: Hull and Vigar (along with others) point to the lack of any 'real' difference between the new Section 54A 'plan-led system' and the erstwhile regime. But planners, developers and the public perceive a difference and have acted accordingly. This has allowed planners to negotiate on the strength of Section 54A when the strength itself is questionable. All this discretion and autonomy has been despite the vastly increased amount and detail of central policy guidance that has aimed to restrict such autonomy. It is clear that significant autonomy existed at the local level and that planners were both willing and able to exercise that discretion often in ways that altered or modified government aims and objectives.

The scope of, and limits to, local discretion are explored by Tewdwr-Jones and Harris. They argue that some rural local authorities, concerned about local socio-economic changes, broke away from the ethos which central government tried to impose. While local planning officers seemed to have retained their respect for the force of government policies (as set out in circulars and PPGs) and the bureaucratic values of consistency between decisions taken at different times, local councillors grew impatient and developed their own criteria (typically, focusing on personal characteristics of applicants for planning permission) to decide applications. The discretion allowed to local planning authorities and the absence of third-party rights of appeal to government against the granting of planning permission, meant that these activities continued for some time without critical scrutiny from outside their localities. However, when such an examination was undertaken then central government intervention to rectify anomalies was direct and (seemingly) effective. Yet, as Tewdwr-Jones and Harris note, the celebrated cases of errant local authorities which they discuss did not come to central government's view as the result of systematic monitoring of, or control over, implementing agencies (i.e. local authorities); and there may well be similar practices being undertaken elsewhere which remain undetected and unpublicised.

Support of interest groups and agencies affected by the policy

As we mentioned with the role of planners above, because of the power and autonomy of groups and individuals charged with implementing policy their support or understanding must be present. It would seem illogical to create scope for discretion by bureaucrats and then imagine that they will not use it. But this appears to be what the New Right forecast would happen: perhaps in the impartial public servants mould of Whitehall, local bureaucrats would also disregard any personal feelings and acquiesce in policy implementation. We have seen the impossibility and unreality of this at an individual level but there are also the wider bodies and groups of interests to consider.

One of the principal characteristics of the New Right and particularly Mrs Thatcher was the stridency of approach which included a lack of compromise or consultation. This applied not only to bodies charged with implementation policy but also to MPs and government departments. The tendency, as demonstrated in the case of SPZs, was to have a small group of interests (usually property-related) who would help develop detailed policies. Where consultation did take place little actually changed. Many of the responses to proposals, particularly regarding deregulation, were unsupported: it would have been like turkeys voting for Christmas in some respects. But other groups and bodies were also unsupportive. Conservation bodies, local amenity societies, countryside protection groups, etc. were all normally consulted on change and were usually opposed to any watering down, by-passing or speeding up of planning controls. It was hardly surprising therefore that local planning authorities and planners could claim another form of local mandate in interpreting and modifying policy. It was not until what was perceived as a 'pro-planning' approach or the sequential test of PPG 6, for example, that implementing agencies began to look more favourably upon central government's proposals.

This picks up some of the differences between the Thatcher and Major governments noted in chapter 1. Major's more conciliatory approach emphasised consultation and compromise as opposed to the 'there is no alternative' view of Mrs Thatcher. But this has not been the only difference. The Major governments have, on the whole, also introduced what could be perceived as more positive approaches to planning. All things being equal we should therefore expect to see a difference in implementation success between the Thatcher and Major Governments. However, there were other important factors that differed and had significant impacts upon the New Right: the changing socio-economic context and political support.

Stable socio-economic contexts which do not undermine political support

As we concluded in chapter 1 the overall aims of the Thatcher and Major governments were similar enough to attach the label New Right to them both, but the mechanisms and emphases did vary. If we crudely identify two periods during the 1980s and 1990s – a broadly anti-planning stance and a broadly pro-planning stance against the backdrop of similar ideological thinking – then we need to identify why these two approaches emerged. Part of the reason comes down to changing circumstances, with the particular watershed being the recession that emerged in the late 1980s. As well as hitting all aspects of the economy the recession undermined the cherished New Right notion of property-led urban regeneration and the role of planning in suppressing demand for economic growth.

It also began to be clear by the late 1980s that the changes introduced by the New Right were not having the desired effect, either because the causal theories were inadequate, as described above, or because there were problems with implementation. It was around this time that other ideas began to emerge, such as the green issues prominent in the European elections of 1989. The individual began to be subsumed by the community and the whole so much so that in 1985 Mrs Thatcher described the Tory Party as 'friends of the earth' (McCormick 1991: 2). But at the same time electoral changes in 1987 and 1992 began to deprive the New Right of the massive majorities they had enjoyed in 1979 and 1983 allowing them to rule essentially without question. Although many of the approaches pursued since 1979 had widespread support more environmentally sensitive issues began to become more important to the electorate.

All of these factors began to undermine the more controversial aspects of the New Right's approach to planning. Although the tactics and mechanisms in many ways altered as Allmendinger and Tewdwr-Jones (1997) have argued, the overall aims have remained consistent. What is evident now more than ever are the contradictions within New Right thinking, not so much between liberalism and authoritarianism but between the market and the environment.

Chapter 1 raised the possibility of a 'locality' effect in the formulation, interpretation and implementation of policy as well as local action and resistance to central policies. As Bagguley *et al.* (1990) suggest, local government has become a focus for locality-specific resistance to change particularly during the 1980s because of the unifying centralisation of Thatcherism. The various chapters in this volume demonstrate that the scope for autonomous local action and the influence of locality-specific approaches to central policy were a significant influence on the policies and success of the New Right approach to planning.

Bramley and Lambert point to the important role of local opposition in defeating the proposed relaxation of Green Belt controls as well as the continued emphasis on 'good' design at the local level and the (perceived) over-use of Section 106 agreements. The perception of an over-concentration of local power was also the case with development plans. Hull and Vigar found developers concerned that the shift to self-adoption gave local authorities *carte blanche* to do what they wished. Similarly, Allmendinger points to the ability of local authorities to interpret policy guidance in ways that were often at odds with the spirit of central government intentions.

The reasons and opportunity for this autonomous local action can be found in a combination of factors, but most important of all was the New Right's 'agency' view of local government and its critique of it as being unrepresentative, financed mainly by central government and a source of opposition to a sovereign parliament (Kavanagh 1990). This resulted in two central visions: local government as an 'implementor' of central government policies and an antagonistic approach to matters such as local finance, control and powers. In reality, a more realistic model would be of a 'partner'. The interpretation of central guidance and legislation at the crude local level combined with the discretion afforded professionals within local government to make day-to-day decisions was enough to allow significant local involvement often in ways that were contrary to central policy direction.

What has happened since 1979?

Andy Thornley makes it clear in his chapter that he is sceptical of the ability of local planning authorities and others to thwart or alter centrally directed policies during the period 1979–97. Using regime theory he points to different levels of control and agenda-setting all of which have been directed by the ghosts of Thatcherism: market orientation and centralisation. The role of local plans is severely constrained by central guidance and their main purpose in any event is to maintain the status quo. Should more radical change be required then special regimes will be set up, e.g. Enterprise Zones or Urban Development Corporations.

Our own interpretation and those of the contributors question this. The eighteen-year period in question witnessed two Prime Ministers and five governments,[1] so there is a limit to the extent of consistency and lessons that can be drawn. As with any government in power, ideology had to be balanced with electoral expediency. We can see this in Mrs Thatcher's much vaunted conversion to green politics following the electoral success of the Green Party in 1989. Nevertheless, there are distinctive and identifiable consistencies running through this period though to view it as homogeneous would be too much; a more appropriate view would be 'similar ends, different means'. To be more specific, we discern two distinct periods (Table 10.1).

There are various reasons for the changes identified in Table 10.1, some of which have been explored by Andy Thornley earlier. We consider that the most important reason is that the government's aims were not achieved in the first period identified (i.e. 1979–1988/89). Working with the existing system and processes and making incremental changes, as we have seen in the contributory chapters, allowed the planning system's inherent flexibility to adapt and to a certain extent overcome the government's intentions. A prime example was design. The government made it clear in Circular 22/80 that design was not something that Local Planning Authorities should concern themselves with. But the only way for applicants to ensure this was through the appeal process. Given the time delays involved, most applicants were naturally wary of this route and the planners knew this. The result was that design was still very much on the agenda in many local authorities regardless of government intentions. In the project-led approach there was little to guide or control this situation. A more 'hands-on' system was therefore needed and introduced where the development plan was the primary 'local' factor determining priorities, etc. The plan would have to take into account central guidance and this would be checked by an Inspector at Inquiry. Rogue decisions and approaches at the local level were therefore to be minimised through the antithesis of New Right planning: the plan itself.

The second reason related to results. We have seen some of these above and in particular the causal assumptions behind, for example, the move towards a market orientation of planning. The whole thrust of the 'planning inhibits development' idea was shown to be simplistic if not plain wrong. Contrary to the government's expectations landowners, developers and householders actually welcomed the certainty provided by the planning system. Planning has clearly become institutionalised into financial, legal, social and political life to such a degree that too much now rests on its being there in essentially its current form. Of course the aim to promote development was still valid, but it was not planning *per se* that was inhibiting it, as Enterprise Zones clearly demonstrated – it was the general economic climate, lack of public-sector infrastructure, lack of demand, etc.

The clearest example of the role of planning in development and redevelopment came through Simplified Planning Zones. Far from clamouring for them very few developers or landowners were interested.

Table 10.1 Changing attitudes towards planning under the New Right

Period	*Approach*	*Means (in order)*	*Ends*	*Examples of change*
1979–1988/1989[2]	'Project-led' Reduce strategic role of planning and make decisions on their merits	1. Market-orientation 2. Rule of law 3. Centralisation	Deregulation of controls (release pent-up demands through targeting supply-side constraints	UDCs EZs SPZs Circular 22/80 'Lifting the Burden' (H.M. Government 1985)
	Development led		Make system more 'transparent' by reducing local planning authorities' discretion Do not alienate conservation-minded voters	
1988/1989–1997	'Plan-led' Reintroduce a (modified) element of strategic role for planning that takes strong central guidance into account (Illusion of) locally led planning to reduce responsibility of Secretary of State	1. Centralisation 2. Rule of law 3. Market orientation	Control planning through strengthened central mechanisms (Grudgingly) introduce environmental concerns and design criteria	Elevation of importance of development plans Revised PPG series Rio commitments Section 54A 'This Common Inheritance' (H.M. Government 1990)

Similarly the intention to remove discretion and introduce a rule of law was elusive. The problem was that local factors were simply too complicated to make any more than simple generalisations about. So, for example, extensions to the General Development and Use Classes Order were fraught with difficulty. The English and Welsh B1 use class is defined as a business that can be carried on without detriment to the amenity of a residential area. Far from reducing discretion planners and others have to become involved in determining 'amenity' and 'detriment'. The result of both inappropriate causal theories *and* lack of progress led the New Right to pursue these matters with less vigour in the second period particularly as they also had the potential to alienate many traditional Conservative voters.

Finally, changing economic, social and political circumstances began to undermine the stridency and purity of the early New Right approaches. The electorate was on the whole not willing to see a wholesale deconstruction of the welfare state including planning. Yes, the system could be more effective and efficient but a bonfire of controls was not a realistic option. The electorate also made their views on the environment more prominent and the government was forced to listen. Europe also became an increasingly important player in UK planning. A number of commentators cite the Environmental Assessment Directive as the first main step down this road and Bullied (1993) claims that the government did not appreciate the full impact it would have on domestic planning policy. Further European involvement has been aided by the Single European Act which permits the European Parliament to introduce environmental protection measures and require their implementation by member states. Recently, the European Commission (EC) has been preparing what is little less than a strategic plan for the whole community – the European Spatial Development Directive (Fyson 1995) – which follows the Europe 2000 document on the potential for future spatial development. In addition, Human (1995) points to the increased EC involvement in tourism and the tendency for cities and regions of the UK to deal with Brussels directly on urban initiatives and by-pass Westminster completely. However, Davies (1996) is less sure about the integration of UK planning within a European system. The more likely outcome, he predicts, is a shift towards the greater use of Environmental Assessment and the introduction of Europe-wide objectives within regional and sub-regional guidance.

Overall, we can point to a changed emphasis in the New Right approach to planning over the period which, given its length, is not altogether unsurprising. There is, however, little doubt that the period witnessed significant changes in planning at both the central and local levels. It would be difficult to argue that some of these changes were not for the better, particularly in the later years with their emphasis on accountability and responsiveness. But other changes were clearly ideologically driven. What is surprising is the extent to which these more ideological changes were thwarted or emasculated by planners and the system itself. And this may be the enduring lesson from the New Right years, particularly for New Labour. The 1980s and 1990s tell us as much about the institutions and practices of planning as about the success or failure of a radical government.

Notes

1 We are counting 1979–83 (Thatcher), 1983–7 (Thatcher), 1990–2 (Major) and 1992–7 (Major).

2 These are rough approximations and one could find examples of legislation and policy changes that contradict the broad thrust of approach. The distinctions are also presented in a rather stark way whereas in reality they were more blurred.

BIBLIOGRAPHY

Adam Smith Institute (1983) *Omega Report: Local Government Policy*, London: Adam Smith Institute.

Adams, D. (1994) *Urban Planning and the Development Process*, London: UCL Press.

Albrechts, L. (1995) 'Innovation in plan making in Belgium', Paper presented to Innovation in Development Plan-Making workshop, Leuven, 26–28 January 1995.

Allmendinger P. (1997) *Thatcherism and Planning. The Case of Simplified Planning Zones*, Aldershot: Avebury.

Allmendinger, P. and Tewdwr-Jones, M. (1997) 'Post-Thatcherite urban planning and politics: a Major change?', *International Journal of Urban and Regional Research*, 21, 1: 100–16.

Ambrose, P. (1986) *Whatever Happened to Planning?*, London: Methuen.

Ambrose, P. (1992) 'Changing planning relations', in P. Cloke (ed.), *Policy and Change in Thatcher's Britain*, Oxford: Pergamon Press.

Amery, C. (1995) 'The ministry of sloth?', *Perspectives on Architecture* 12: 26–9.

Anderson, J. (1983) 'Geography as ideology and the politics of crisis: the EZ experiment', in J. Anderson, S. Duncan and B. Hudson, *Redundant Spaces in Cities and Regions*, London: Academic Press.

Andreae, S. (1996) 'From comprehensive development to conservation areas', in M. Hunter (ed.), *Preserving the Past: the Rise of Heritage in Modern Britain*, Stroud: Sutton.

Anstey, J. (1993) Letter to the Editor, *The Times*, 29 March.

Ashworth, G.J. (1992) 'Heritage and tourism: an argument, two problems and three solutions', in C.A.M. Fleischer van Rooijen (ed.), *Spatial Implications of Tourism*, Groningen: GeoPers.

Ashworth, G.J. and Tunbridge, J.E. (1990) *The Tourist-Historic City*, London: Belhaven Press.

Ashworth, G.J. and Voogt, H. (1990), *Selling the City*, London: Belhaven Press.

Association of District Councils (1984) Letter to DoE, 20 August.

Atkinson, R. and Moon, G. (1994) *Urban Policy in Britain*, London: Macmillan.

Audit Commission (1989) *Urban Regeneration and Economic Development*, London: HMSO.

Audit Commission (1992) *Building in Quality: a Study of Development Control*, London: HMSO.

Bagguley, P., Mark-Lawson, J., Shapiro, D., Urry, J., Walby, S. and Warde, A. (1990) *Restructuring Place, Class and Gender*, London: Sage.

Bain, C., Dodd, A. and Pritchard, D. (1990) *RSPB Planscan: a Study of Development Plans in England and Wales*, Conservation Topic Paper No. 28, Sandy: RSPB.

Baldock, D. (1989) 'The European Community and conservation in the Thatcher decade', *ECOS*, 10, 4: 33–7.

Banham, R., Barker, P., Hall, P. and Price, C. (1969) 'Non-plan: an experiment in freedom', *New Society*, 20 March.

Bar-Hillel, M. (1991) 'Conservationists' nightmare or developers' dream?', *Chartered Surveyor Weekly*, 14 March: 22.

Barlow, J. (1995) 'The politics of urban growth: "boosterism" and "nimbyism" in European boom regions', *International Journal of Urban and Regional Research* 19, 1: 129–44.

Barlow, J. and Chambers, D. (1992) 'Planning agreements and social housing quotas', *Town and Country Planning*, 61; 5: 136–42.

Barlow, J., Cocks, R. and Parker, M. (1994) *Planning for Affordable Housing*, DoE Planning Research Report, London: HMSO.

Barr, C., Howard, D., Bunce, B., Gillespie, M. and Hallam, C. (1991) *Changes in Hedgerows in Britain between 1984 and 1990*, Grange-over-Sands: Institute for Terrestrial Ecology.

Barrett, H.J. (1996) 'Townscape change and local planning management in city centre conservation areas: the example of Birmingham and Bristol', unpublished PhD thesis, Birmingham: School of Geography, University of Birmingham.

Barrett, H.J. and Larkham, P.J. (1994) *Disguising Development: Façadism in City Centres*, Research Paper 11, Birmingham: Faculty of the Built Environment, University of Central England.

Barrett, S. and Fudge, C. (eds) (1981) *Policy and Action: Essays on Implementation of Public Policy*, London: Methuen.

Barton, H. (1994) 'Research on the environmental appraisal of development plans', in E. Wilson (ed.), *Issues in the Environmental Appraisal of Development Plans*, Working Paper No. 153, Oxford: Oxford Brookes University School of Planning.

Barton, H. (1995) 'The capacity to deceive', *ECOS*, 16, 3/4: 18–25.

Bateman, M. (1985) *Office Development: a Geographical Analysis*, London: Croom Helm.

Beckerman, W. (1995) *Small is Stupid: Blowing the Whistle on the Greens*, London: Duckworth.

Bedfordshire County Council (1995) *Bedfordshire structure plan 2011: deposit draft*, Bedford: Bedfordshire County Council.

Bedfordshire County Council and RSPB (1996) *A step by step guide to environmental appraisal*, Bedford: Bedfordshire County Council.

Bell, P. and Cloke, P. (1989) 'The changing relationship between the private and public sectors: privatisation and rural Britain', *Journal of Rural Studies*, 5, 1: 1–15.

Benson, J.K. (1983) 'Interorganizational networks and policy sectors', in D. Rogers and D. Whetton (eds), *Interorganizational Organisation*, Ames: Iowa State University Press.

Best, R.H. and Rogers, A.W. (1973) *The Urban Countryside: the Land Use Structure of Small Towns and Villages in England and Wales*, London: Faber and Faber.

Biggs, S. and Travers, T. (1994) 'Opportunities for city-wide government in

London: the experience of the metropolitan areas', *Local Government Policy-Making*, 21, 2.

Binney, M. (1991) 'Guardians of England's glory demeaned', *The Times*, 16 April.

Birmingham City Council (1984) *Birmingham central area local plan*, Birmingham: Birmingham City Council.

Birmingham City Council (1986) *Conservation in the environment: a strategy for Birmingham*, unpublished draft report, Birmingham: Development Department, Birmingham City Council.

Birmingham City Council (1987) *Conservation strategy*, first edition, Birmingham: Birmingham City Council.

Birmingham Conservation Areas Advisory Committee (1984) *Development pressures in the city centre*, unpublished discussion paper, Birmingham: Conservation Areas Advisory Committee.

Bishop, K.D. (1997) 'The challenge of convergence', in R. Macdonald and H. Thomas (eds), *Nationality and Planning in Scotland and Wales*, Cardiff: University of Wales Press.

Bishop, K.D., Philips, A.A.C. and Warren, L.M. (1995) *Protected Areas in Wales*, Bangor: Countryside Council for Wales.

Bishop, K.D. and Phillips, A.A.C. (1993) 'Seven steps to market – the development of the market-led approach to countryside conservation and recreation', *Journal of Rural Studies*, 9, 4: 315–38.

Blackhall, C. and Graham, D. (1996) 'Planning for progress in development', *Planning Week*, 25 April.

Boddy, M. (1983) 'Local economic and employment strategies', in M. Boddy and C. Fudge (eds), *Local Socialism?*, London: Macmillan.

Boddy, M., Lambert, C. and Snape, D. (1997) *A City for the 21st Century? Globalization, Planning and Urban Change in contemporary Britain*, Bristol: Policy Press.

Bond, R. (1988) 'Funding the finds', *Surveyor*, 4997: 10–13.

Booth, E. (1993) 'Enhancement in conservation areas', *The Planner*, 79, 4: 22–3.

Booth, P. (1996) *Controlling Development: Certainty and Discretion in Europe, the USA and Hong Kong*, London: UCL Press.

Bradshaw, J. (1992) 'Social Security', in D. Marsh and R.A.W. Rhodes (eds), *Implementing Thatcherite Policy. Audit of an Era*, Buckingham: Open University Press.

Bramley, G. and Watkins, C. (1995) *Circular projections: household growth, housing development and the household projections*, London: CPRE.

Bramley, G. and Watkins, C. (1996) *Steering the Housing Market: New Building and the Changing Planning System*, Bristol: Policy Press.

Bramley, G., Bartlett, W. and Lambert, C (1995) *Planning, the Market and Private Housebuilding*, London: UCL Press.

Brecon Beacons National Park (1994) *Brecon Beacons National Park local plan consultation draft*, Brecon: Brecon Beacons National Park.

Breheny, M. and Hall, P. (eds) (1996) *The people – where will they go?* National report of the TCPA regional inquiry into housing need and provision in England. London: Town and Country Planning Association.

Brindley, T., Rydin, Y. and Stoker, G. (1996) *Remaking Planning: the Politics of Urban Change in the Thatcher Years*, 2nd ed., London: Unwin Hyman.

Bristol City Council (1976) *Bristol city docks local plan: draft district plan report*, Bristol: Bristol City Council.

Bristol City Council (1984) *Conservation policies*, draft document, Bristol: Bristol City Council, City Planning Department.

Bristol City Council (1987) *City centre local plan: 16–20 Narrow Quay planning brief*, Bristol: Bristol City Council, City Planning Department.

Bristol City Council (1989) *Conservation policies*, Bristol: Bristol City Council, Urban Design Section and Development Control Division, City Planning Department.

Bristol City Council (1990) *Bristol city centre Local Plan: draft written statement*, Bristol: Bristol City Council.

Brownill, S. (1990) *Developing London Docklands. Another Great Planning Disaster?*, London: Paul Chapman.

Brownill, S. and Thomas, H. (1993) 'The Docklands experience: locality and community in London', in R. Imrie and H. Thomas (eds), *British Urban Policy and the Urban Development Corporations*, London: Paul Chapman.

Brownill, S. (1997) 'Local governance and the racialisation of urban policy in the UK: the case of urban development corporations', *Urban Studies*, 33, 8, 1337–55.

Brunskill, R.W. (1993) 'The achievements of the accelerated resurvey of 1982–7', in M. Robertson (ed.), 'Listed buildings: the national resurvey of England', *Transactions of the Ancient Monuments Society* 37: 21–94.

Bruton, M. and Nicholson, D. (1987) *Local Planning in Practice*, London: Hutchinson.

Buchanan, J. (1978) *The economics of politics*, IEA Occasional Paper. London: Institute of Economic Affairs.

Buchanan, J. and Tullock, G. (1962) *The Calculus of Consent*, Ann Arbor: University of Michigan Press.

Buckley, P. (1995) 'Critical natural capital: operational flaws in a valid concept', *ECOS* 16, 3/4: 12–18.

Bullied, P. (1993) 'Coded assessment message garbled in translation', *Planner*, 1042.

Bulpitt, J. (1986) 'The Thatcher statecraft', *Political Studies*, 14, 6.

Burke, T. (1995) 'View from the inside: UK environmental policy seen from a practitioner's perspective', in T.S. Gray (ed.), *UK Environmental Policy in the 1990s*, Basingstoke: Macmillan.

Burleigh, D. (1993) 'Integrating environmental concerns into European Community regional policy', unpublished M.Sc. thesis, Oxford Brookes University.

Burton, I. and Drewry, G. (1990) 'Public legislation: a survey of the sessions 1983/1984 and 1984/1985', *Parliamentary Affairs*, 41.

Butterfly Conservation, Friends of the Earth, Plantlife, The Royal Society for Nature Conservation – the Wildlife Trusts' partnership, The Royal Society for the Protection of Birds and World Wide Fund for Nature (1993) *Biodiversity Challenge: an Agenda for Conservation Action in the UK*, Sandy: The Royal Society for the Protection of Birds.

Cameron-Blackhall, J. (1993) *The performance of simplified planning zones*, Department of Town and Country Planning, University of Newcastle upon Tyne Working Paper No. 30.

Carter, N. and Lowe, P. (1995) 'The establishment of a cross-sector environment agency', in T.S. Gray (ed.), *UK Environmental Policy in the 1990s*, Basingstoke: Macmillan.

CBI (1991) *A London development agency: optimising the capital's assets*, London: Confederation of British Industry.

Champion, A.G. and Townsend, A.R. (1991) *Contemporary Britain: a Geographical Perspective*, London: Hutchinson.

Cheshire, P. and Sheppard, S. (1989) 'British planning policy and access to housing: some empirical estimates', *Urban Studies*, 26: 469–85.

Chester City Council (1996) *Chester district local plan: first draft for public consultation*, Chester: Chester City Council Planning Services.

Clark, D. (1986) 'Double standards at the National Trust', *The Observer*, 24 August: 8.

Clark, G., Darral, J., Grove-White, R., McNaughton, P. and Urry, J. (1994) *Leisure Landscapes*, London: Council for the Protection of Rural England.

Cleethorpes Borough Council (1987) Draft Local Plan, Cleethorpes.

Cleethorpes Borough Council (1988) North Promenade simplified planning zone.

Cleethorpes Borough Council (1988a) Draft Local Plan – second draft, Cleethorpes.

Cloke, P. (ed.) (1992) *Policy and Change in Thatcher's Britain*, Oxford: Pergamon Press.

Cloke, P. (1992a) 'The countryside: development, conservation and an increasingly marketable commodity', in P. Cloke (ed.), *Policy and Change in Thatcher's Britain*, Oxford: Pergamon Press.

Cloke, P. (1996) 'Housing in the open countryside: windows on irresponsible planning in rural Wales,' *Town Planning Review*, 67, 3.

Cloke, P. and Goodwin M. (1992) 'Conceptualising countryside change: from post-Fordism to rural structured coherence', *Transactions of the Institute of British Geographers*, 17: 321–36.

Cloke, P. and Little, J. (1990) *The Rural Local State*, Oxford: Clarendon Press.

Cloke, P. and McLaughlin, B. (1989) 'Politics of the Alternative Land Use and Rural Economy (ALURE) proposals in the UK: Crossroads or blind alley? *Land Use Policy* 6, 3: 235–48.

Clover, C. (1987) 'Minister rejects fear of rural building bonanza', *Daily Telegraph*, 20 February.

Commission of the European Communities (1985) *Assessment of the effects of certain public and private projects on the environment*, Directive 85/337/EEC, Luxembourg: Office for Official Publications of the European Communities.

Commission of the European Communities (1990) *Freedom of access to information on the environment*, Directive 90/313/EEC, Luxembourg: Office for Official Publications of the European Communities.

Commission of the European Communities (1992) *Towards sustainability: a European Community programme of policy and action in relation to the environment and sustainable development*, COM(92)23 Final, Luxembourg: Office for Official Publications of the European Communities.

Commission of the European Communities (1994) *Interim review of implementation of the European Community programme of policy and action in relation to the environment and sustainable development, towards sustainability*, Luxembourg: Office for Official Publications of the European Communities.

Conservative Research Department (1995) *Rural White Paper: Conservative research briefing*, London: Conservative Research Department.

Cooke, P. (1989) 'Restructuring, flexibility and local labour markets', in J. Morris (ed.), *Labour Market Responses to Industrial Restructuring and Technological Change*, Brighton: Wheatsheaf.

Coopers and Lybrand, Deloitte, (1991) *'London: world city moving into the 21st century'*, London: HMSO.

Council for the Protection of Rural England (1990) *Building responsibilities: the case for extending planning control over agricultural and forestry buildings*, London: Council for the Protection of Rural England.

Country Landowners Association (1995) *Towards a rural policy: a vision for the 21st century*, London: Country Landowners Association.

Countryside Commission (1981) *Countryside commission: thirteenth annual report, 1979–80*, London: HMSO.

Countryside Commission and English Nature (1997) *The character of England: landscape, wildlife and natural features*, Cheltenham: Countryside Commission.

Crewe, I. (1994) 'Electoral behaviour', in D. Kavanagh and A. Seldon (eds), *The Major Effect*, London: Macmillan.

Cullingworth, B. and Nadin, V. (1994) *Town and Country Planning in Britain*, 11th edition, London: Routledge.

Curry, N.R. (1994) *Countryside Recreation, Access and Land Use Planning*, London: Spon.

Curtis, C. and Headicar, P. (1994) *The location of new residential development: its impact on car-based travel – 1. research design and methodology*, Working Paper No. 154, Oxford: Oxford Brookes University School of Planning.

Dabinett, G. and Ramsden, P. (1993) 'An urban policy for people: lessons from Sheffield', in R. Imrie and H. Thomas (eds), *British Urban Policy and the Urban Development Corporations*, London: Paul Chapman.

Davies C., Pritchard, D., and Austin, L. (1992) *RSPB PlanScan Scotland: a study of development plans in Scotland*, Sandy: RSPB.

Davies, H.W.E. (1996) 'Planning and the European question', in M. Tewdwr-Jones (ed.), *British Planning Policy in Transition*, London: UCL Press.

Davoudi, S. and Healey, P. (1995) 'City challenge – a sustainable mechanism or temporary gesture', in R. Hambleton and H. Thomas (eds), *Urban Policy Evaluation*, London: Paul Chapman.

Delafons, J. (1993) 'The conservation see-saw', *Town and Country Planning*, 62, 9: 226–8.

Delafons, J. (1995) 'Do we trust the National Trust?', *Town and Country Planning*, 64, 5/6: 150–1.

Department of National Heritage (1992) *Circular 1/92: Responsibilities for conservation policy and casework*, London: HMSO.

Department of the Environment and Welsh Office (1980) *Circular 22/80: Development control: policy and practice*, London: HMSO.

Department of the Environment (1980a) *Circular 9/80: Land for private housebuilding*, London: HMSO.

Department of the Environment (1981) *Local Government Planning and Land Act 1980 – Circular 23/81*, London: HMSO.

Department of the Environment (1983) *Local development schemes*, Internal Paper.

Department of the Environment (1984) *Circular 14/84: Green Belts*, London: HMSO.

Department of the Environment (1984a) *Circular 9/84: Planning controls over hazardous development*, London: HMSO.

Department of the Environment and Welsh Office (1984b) *Circular 16/84: Industrial development*, London: HMSO.

Department of the Environment (1984c) *Circular 22/84: Memorandum on structure and local plans*, London: HMSO.

Department of the Environment (1984d) *Simplified planning zones*, consultation paper, London: HMSO.

Department of the Environment (1985) *Circular 14/85: Development and employment*, London: HMSO.

Department of the Environment (1985a) *Circular 31/85: Aesthetic control*, London: HMSO.

Department of the Environment (1985b) Letter to Slough Estates, 14th February.

Department of the Environment (1985c) *Circular 1/85: The use of conditions in planning permissions*, London: HMSO.

Department of the Environment, (1985d) *Lifting the Burden*, London: HMSO.

Department of the Environment (1986) *Circular 2/86: Development by small businesses*, London: HMSO.

Department of the Environment (1986a) *The future of development plans*, London: HMSO.

Department of the Environment (1987) *Town and country planning use classes order 1987*, London: HMSO.

Department of the Environment and Welsh Office (1987a) *Circular 16/87: Development involving agricultural land*, London: HMSO.

Department of the Environment (1988) *Planning policy guidance note 5: Simplified planning zones*, London: HMSO.

Department of the Environment (1988a) *Releasing enterprise*, London: HMSO.

Department of the Environment and Welsh Office (1988b) *Planning policy guidance note 7: Rural enterprise and development*, London: HMSO.

Department of the Environment (1989) *Permitted use rights in the countryside*, consultation paper, London: Department of the Environment.

Department of the Environment (1989a) *The future of development plans*, Cm. 569. London: HMSO.

Department of the Environment (1990) *This common inheritance: Britain's environmental strategy*, London: HMSO.

Department of the Environment (1991) *Development control statistics, England 1990–91*, London: HMSO.

Department of the Environment (1991a) *Policy appraisal and the environment*, London: HMSO.

Department of the Environment (1991b) *Simplified planning zones: progress and procedures*, London: HMSO.

Department of the Environment (1992) *Planning policy guidance note 3: Housing*, London: HMSO.

Department of the Environment (1992a) *Climate change: our national programme for CO_2 reductions*, London: Department of the Environment.

Department of the Environment (1992b) *Circular 11/92: Planning controls for hazardous substances; the Planning (Hazardous Substances) Act 1990; the Planning (Hazardous Substances) Regulations 1992 (SI 1992 No. 656)*, London: HMSO.

Department of the Environment (1992c) *Planning, pollution and waste management*, London: HMSO.

Department of the Environment (1992d) *The UK environment*, London: HMSO.

Department of the Environment (1992e) *Planning policy guidance note 12: Development plans and regional planning guidance*, London: HMSO.

Department of the Environment and Welsh Office (1992f) *Planning policy guidance note 7: The countryside and the rural economy*, London: HMSO.

Department of the Environment (1992g) *Planning policy guidance note 1: General policy and principles*, London, HMSO.

Department of the Environment (1993) *Environmental appraisal of development plans: a good practice guide*, London: HMSO.

Department of the Environment (1993a) *Enquiry into the planning system in North Cornwall district*, London: HMSO.

Department of the Environment, (1993b) *Planning policy guidance note 6: Town centres and retail developments*, London: HMSO.

Department of the Environment and Department of Transport (1993c) *Reducing transport emissions through planning*, London: HMSO.

Department of the Environment (1993d) *London – making the best better*, London: HMSO.

Department of the Environment (1993e) News Release: 'Local Authorities urged to make positive plans for the future', 6 Jan. 1993, London: DoE.

Department of the Environment and Department of Transport (1994) *Planning policy guidance note 13: Transport*, London: HMSO.

Department of the Environment (1994a) *The costs of determining planning applications and the development control service*, London: HMSO.

Department of the Environment (1994b) *Planning policy guidance note 23: Planning and pollution control*, London: HMSO.

Department of the Environment (1994c) *Planning policy guidance note 9: Nature conservation*, London: HMSO.

Department of the Environment and Department of National Heritage (1994d) *Planning policy guidance note 15: Planning and the historic environment*, London: HMSO.

Department of the Environment (1994e) *Improving the local plan process*, London: DoE.

Department of the Environment (1994f) *Regional planning guidance note 9: Regional planning guidance for the South East*, London: HMSO.

Department of the Environment (1995) *Making waste work: a sustainable waste strategy*, London: Department of the Environment.

Department of the Environment (1995a) *Projections of households in England to 2016*, London: HMSO.

Department of the Environment (1995b) *Town centres and retail developments*: Draft PPG6, London: HMSO.

Department of the Environment and Ministry of Agriculture, Fisheries and Food (1995c) *Rural England*, London: HMSO.

Department of the Environment (1995d) *London gets voice closer to government*, Press Release 575, November 23rd.

Department of the Environment (1996) *Household growth: where shall we live?* London: HMSO.

Department of the Environment (1996a) *Changes in the quality of environmental statements for planning projects*, London: HMSO.

Department of the Environment (1996b) *Revision of Planning policy guidance note 23: Waste*, London: Department of the Environment.

Department of the Environment and Scottish Office (1996c) *The United Kingdom air quality strategy: consultation draft*, London: Department of the Environment.

Department of the Environment (1997) *Indicators of sustainable development for the United Kingdom*, London: HMSO.

Department of the Environment (1997a) *Planning policy guidance note 7: The countryside – environmental quality and economic and social development*, London: HMSO.

Derby City Council (1985) Local plan for Rosehill and Peartree, Derby.

Devon County Council (1995) *Devon county structure plan Devon 2011: first review consultation draft*, Exeter: Devon County Council.

DiGaetano, A. and Klemanski, J. (1993) 'Urban regimes in comparative perspective', *Urban Affairs Quarterly*, 29, 1.

Dinefwr Borough Council (1995) *Dinefwr local plan draft deposit part 1: written statement*, Llandeilo, Dyfed: Dinefwr Borough Council.

Discipline Network in Town Planning (1996) *Annual Report 1994–95*, London: University of Westminster Press.

Dodd A. and Pritchard D. (1993) *PlanScan Northern Ireland: a study of development plans in Northern Ireland*, Sandy: RSPB.

Downs, A. (1957) *An Economic Theory of Democracy*, New York: Harper and Row.

Downs, A. (1967) *Inside Bureaucracy*, Boston: Little Brown.

Duncan, S. and Goodwin, M. (1988) *The Local State and Uneven Development*, London: Polity Press.

Dunsire, A. (1978) *Implementation in a Bureaucracy*, Oxford: Martin Robertson.

Edgar, E. (1983) 'Bitter harvest', *New Socialist*, Sept/Oct.

Edwards, M. and MacCafferty, A. (1992) '1991: a time to reflect', *Estates Gazette*, No. 9223.

EIU (Economist Intelligence Unit) (1975) *Housing land availability in South East England*, London: HMSO.

Elmore, R. (1982) 'Backward mapping', in W. Williams (ed.), *Studying Implementation*, New York: Chatham House.

Elson, M. (1986) *Green Belts: Conflict mediation in the Urban Fringe*, London: Heinemann.

Elson, M., MacDonald, R. and Steenberg, C. (1995) *Planning for rural diversification*, London: HMSO.

English Heritage (1992) *Managing England's heritage: setting our priorities for the 1990s*, London: English Heritage.

English Nature (1994) *Sustainability in practice*, Peterborough: English Nature.

Entec UK (1996) 'The application of environmental capacity to land-use planning', Final report to the Department of the Environment, Leamington: Entec UK.

Environment Agency (1996) *River Thames (Buscot to Eynsham), Windrush and*

Evenlode Local Environment Agency Plan: consultation draft, Wallingford: Environment Agency Thames Region West Area.

Environmental Data Services (1996) 'NRA staff dominate top posts in Environment Agency', *The ENDS report*, 247: 3.

European Environment Agency (1995) *Environment in the European Union: report for the review of the fifth environmental action programme*, Luxembourg: Office for Official Publications of the European Communities.

Evans, A. (1987) *House prices and land prices in the South East: a review*, London: Housebuilders Federation.

Evans, A. (1991) 'Rabbit hutches on postage stamps: planning development and political economy', *Urban Studies*, 28: 6, 853–70.

Ezard, J. (1995) 'Fings ain't what they used to be at Fawlty Towers', *The Guardian*, 9 February: 22.

Fainstein, N. and Fainstein, S. (1982) 'Restoration and struggle: urban policy and social forces', in N. Fainstein and S. Fainstein (eds), *Urban Policy under Capitalism*, Newbury Park, Calif.: Sage.

Farthing, S. (1996) 'Planning and social housing provision', in C. Greed (ed.), *Investigating Town Planning*, London: Longman.

Farthing, S., Coombs, T. and Winter, J. (1993) 'Large development sites and affordable housing', *Housing and Planning Review*, 48: 1, 11–13.

Farthing, S., Malpass, P. and Lambert, C. (1996) *Land, planning and housing associations*, London: Housing Corporation Research Series.

Fife Regional Council (1995) *Sustainability Indicators for Fife*, Glenrothes: Fife Department of Economic Development and Planning.

Fishlock, M. (1992) *The Great Fire at Hampton Court*, London: Herbert Press.

Fowler, P. (1989) 'Heritage: a post-modern perspective', in D.L. Uzzell (ed.), *Heritage Interpretation: the Natural and Built Environment*, London: Belhaven.

Franks, A. (1988) 'The street they froze in time', *The Times*, 15 July: 11.

Fyson, A. (1995) 'Europe at a distance', *Planning Week*, Vol. 3.

Gamble, A. (1984) 'This lady's not for turning: Thatcherism mk III', *Marxism Today*, July.

Gamble, A. (1988) *The Free Economy and the Strong State*, London: Macmillan.

Gardiner, J. (1994) 'The environmental appraisal of development plans: a view from the NRA as a statutory consultee authority', in E. Wilson (ed.) *Issues in the Environmental Appraisal of Development Plans*, Working Paper No. 153, Oxford: Oxford Brookes University School of Planning.

Gatenby, I. and Williams, C. (1996) 'Interpreting planning law', in M. Tewdwr-Jones (ed.), *British Planning Policy in Transition*, London: UCL Press.

Gaze, J. (1988) *Figures in a Landscape: a History of the National Trust*, London: Barrie and Jenkins.

Gold, J.R. and Ward, S.V. (1994) *Place Promotion*, Chichester: Wiley.

Goodchild, B. (1990) 'Planning and the modern/postmodern debate', *Town Planning Review* 60: 119–37.

Goverment Office for London (1996) Strategic Guidance for London Authorities, RPG3, London: Government Office for London.

Grant, M. (1990) *Urban Planning Law*, London: Sweet and Maxwell.

Grant, M. (1991) 'Recent developments', *Encyclopedia of Planning Law and Practice Monthly Bulletin*, July.

Grant, M. (1992) 'Planning law and the British land-use planning system: an overview', *Town Planning Review*, 63: 1, 3–12.

Gray, J. (1993) *Beyond the New Right: Markets, Government and the Common Environment*, London: Routledge.

Gray, T.S. (1995) *UK Environmental Policy in the 1990s*, Basingstoke: Macmillan.

Greenwell, E. (ed.) (1989) *Enterprise and the Rural Economy*, London: Country Landowners Association.

Griffiths, R. (1986) 'Planning in retreat? Town planning and the market in the 1980s', *Planning Practice and Research*, 1.

Grigson, W. (1995) *The Limits of Environmental Capacity*, London: House Builders Federation.

Grimley J.R. Eve (1992) *The use of planning agreements*, DoE Planning Research Report, London: HMSO.

Haigh, N. (1992 *et seq.*) *Manual of Environmental Policy: The EC and Britain*, Harlow: Longman in association with Institute for European Environmental Policy.

Haigh, N. and Lanigan, C. (1995) 'Impact of the European Union on UK environmental policy-making', in T.S. Gray (ed.), *UK Environmental Policy in the 1990s*, Basingstoke: Macmillan.

Hakim, C. (1987) *Research Design. Strategies and Choices in the Design of Social Research*, London: Unwin Hyman.

Hall, P. (1977) 'The inner cities dilemma', *New Society*, 3.

Hall, P. (1984) 'Enterprises of great pith and moment?', *Town and Country Planning*, 53, 11.

Hall, P. (1995) 'Towards a general urban theory', in J. Brotchie *et al.* (eds), *Cities in Competition*, Melbourne: Longman.

Hall, P., Gracey, H., Drewett, R. and Thomas, R. (1973) *The Containment of Urban England*, London: Allen and Unwin.

Hall, S. and Jacques, M. (eds) (1983) *The Politics of Thatcherism*, London: Lawrence and Wishart.

Ham, C. and Hill. M. (1993) *The Policy Process in the Modern Capitalist State*, second edition, Hemel Hempstead: Harvester Wheatsheaf.

Harding, A. (1994) 'Urban regimes and growth machines: towards a cross-national agenda', *Urban Affairs Quarterly*, 29, 3.

Harding, A. (1995) 'European city regimes? Inter-urban competition in the new Europe', paper to the ESRC Local Governance Programme Conference, Exeter, September.

Harrison, A.J. (1977) *Economics and Land-Use Planning*, London: Croom Helm.

Harvey, D. (1989) *The Urban Experience*, Oxford: Basil Blackwell.

Harvey, J.H. (1993) 'The origin of listed buildings', *Transactions of the Ancient Monuments Society*, 37: 1–20.

Headicar, P. (1995) 'PPG 13 – can the honeymoon begin without a bride?', *Oxford Planning Monographs*, 1, 2. Oxford: Oxford Brookes University.

Heald, D.A. (1984) 'Privatisation: analysing its appeal and limitations', *Fiscal Studies*, 5: 36–46.

Healey, P. (1983) *Local Plans and British Land Use Planning*, London: Pergamon.

Healey, P. (1989) 'Directions for change in the British planning system', *Town Planning Review*, 60: 125–49.

Healey, P. (1992) 'The reorganisation of state and market in planning', *Urban Studies*, 29: 411–34
Healey, P. (1993) 'The reorganisation of state and market in planning', in R. Paddison, B. Lever and J. Money (eds), *International Perspectives in Urban Studies*, London: Jessica Kingsley Publishers.
Healey, P. and Shaw, T. (1993) 'Planners, plans and sustainable development', *Regional Studies*, 27, 8: 769–76.
Healey, P., Doak, A., McNamara, P. and Elson, M. (1988) *Land Use Planning and the Mediation of Urban Change: the British Planning System in Practice*, Cambridge: Cambridge University Press.
Healey, P., Davoudi, S., O'Toole, M., Tavsanoglu, S. and Usher, D. (eds) (1992) *Rebuilding the City*, London: Spon.
Healey, P., Purdue, M. and Ennis, F. (1993) *Gains from Planning? Dealing with the Impacts of Development*, York: Joseph Rowntree Foundation.
Hebbert, M. (1992) 'Environmental foundation for a new kind of town and country planning', *Town and Country Planning*, 61, 6: 166–8.
Hebbert, M. and Travers, T. (1988) *The London Government Handbook*, London: Cassell.
Herbert, D.T. (ed.) (1995) *Heritage, Tourism and Society*, London: Mansell.
Herbert-Young, N. (1995) 'Reflections on Section 54A and "Plan-led" decision-making', *Journal of Planning and Environmental Law*, B33–B44, 292–305.
Hertfordshire County Council (1994) *Hertfordshire County Structure Plan Review: Future Directions. Environmental Appraisal Stage 1: Consultation Draft*, Hertford: Hertfordshire County Council.
Hewison, R. (1987) *The Heritage Industry: Britain in a State of Decline*, London: Methuen.
Hewison, R. (1993) 'State and the arts: department of national heritage', *The Times*, 23 May.
Hewison, R. (1995) *Culture and Consensus*, London: Methuen.
Hills, J. (ed.) (1996) *New Inequalities: the Changing Distribution of Income and Wealth in the United Kingdom*, Cambridge: Cambridge University Press.
Hirst, C. (1996) 'An avenging angel of the Tory right . . .', *Planning Week*, 4: 12–13.
Hirst, P. (1989) *After Thatcher*, London: Collins.
Hjern, B. and Hull, C. (1982), 'Implementation research as empirical constitutionalism', in B. Hjern and C. Hull (eds), Implementation beyond hierarchy, *European Journal of Political Research*.
H.M. Government (1991) *Open Government*, London: HMSO.
H.M. Government (1994) *Competitiveness*, London: HMSO.
H.M. Government (1994a) *Biodiversity: The UK Action Plan*, London: HMSO.
H.M. Government (1994b) *Sustainable Development: the UK Strategy Plan*, London: HMSO.
H.M. Government (1994c) *Climate Change: The UK Programme*, London: HMSO.
H.M. Government (1994d) *Sustainable Forestry: The UK Programme*, London: HMSO.
H.M. Government (1995) *Rural England: A Nation Committed to a Living Countryside*, London: HMSO.

H.M. Inspectorate of Pollution (1995) *Planning Liaison with Local Authorities*, London: HMIP.

Hjern, B., and Porter, D.O. (1981) 'Implementation structures. A new unit of administrative analysis', in M. Hill (ed.) (1993) *The Policy Process: A Reader*, London: Harvester Wheatsheaf.

Hogwood, B.W. and Gunn, L.A. (1984) *Policy Analysis for the Real World*, Oxford: Oxford University Press.

Holliday, I. (1991) 'The new suburban right in British local government – Conservative views of the local, *Local Government Studies*, November/December, 45–62.

Holliday, I. (1993) 'Organised interests after Thatcher', in P. Dunleavy *et al.* *Developments in British Politics*, Basingstoke: Macmillan.

Holliday, I., Marcou, C. and Vickerman, R. (1991) *The Channel Tunnel, Public Policy, Regional Development, and European Integration*, London: Belhaven.

Holmans, A. (1995) *Housing Demand and Need in England 1991–2011*, York: York Publishing Services for the Joseph Rowntree Foundation.

Home, R. (1987) *Planning Use Classes*, Oxford: Blackwell Scientific.

Hood, C.C. (1976) *The Limits of Administration*, London: John Wiley.

House of Commons (1978) Third Report from the Expenditure Committee, *National Land Fund*, London: HMSO.

House of Commons (1994) Third Report of the National Heritage Committee, *Our heritage: preserving it, prospering from it*, London: HMSO.

House of Commons (1995) First Report of the Defence Committee, *The defence estate*, London: HMSO.

House of Commons Welsh Affairs Committee (1993) *Rural housing*, London: HMSO.

Howarth, R.W. (1985) *Farming for Farmers?*, London: Institute of Economic Affairs.

Hull, A.D., Marvin, S. and de Cani, R. (1994) *Renewable Energy Policies and Development Plans: A Review of 457 Local Planning Authorities in England and Wales*, ETSU, Didcot.

Hull, A.D., Healey, P. and Davoudi, S. (1995) *Greening the Red Rose County: Working towards an integrated sub-regional strategy*, Department of Town and Country Planning WP No. 58, University of Newcastle.

Human, B. (1995) 'Europe takes a broader view of tourism', *Planner*, 9 June.

Imrie, R. and Thomas, H. (eds) (1993) *British Urban Policy and the Urban Development Corporations*, London: Paul Chapman.

Jackson, P. (1982) *The Political Economy of Bureaucracy,* Oxford: Phillip Allen.

Jackson, P. (1992) 'Economic Policy', in D. Marsh and R.A.W. Rhodes (eds), *Implementing Thatcherite policy. Audit of an Era*, Buckingham: Open University Press.

Jacobs, M. (1993) *Sense and Sustainability*, London: CPRE.

Jenkin, P. (1984) 'Secretary of State's Address to RTPI Summer School', *The Planner*, February.

Jenkins, J. and James, P. (1994) *From Acorn to Oak Tree: the Growth of the National Trust 1895–1994*, London: Macmillan.

Job, S. (1984) 'From EZ to SPZ: lessons from the enterprise zones', Town and Country Planning Association conference paper.

Johnson, C. (1991) *The Economy under Mrs Thatcher 1979–1990*, London: Penguin.

Johnson, S. (1996) 'Repair grants down by nearly £7 million', *Conservation Bulletin* 30: 14–17.

Jones, A. (1996) Local planning policy – the Newbury approach, in M. Tewdwr-Jones (ed.) *British Planning Policy in Transition*, London: UCL Press.

Jones, A.N. and Larkham, P.J. (1993) *The character of conservation areas*, London: Royal Town Planning Institute.

Joseph Rowntree Foundation (1994) *Inquiry into planning for housing*. York: Joseph Rowntree Foundation.

Journal of Planning and Environment Law (1991) 'Current topics: Scheduled Monument Consent', *Journal of Planning and Environment Law*, April: 301.

Journal of Planning and Environment Law (1995) 'R. v. Secretary of State for the Environment, ex p. Mayor and Burgesses of the London Borough of Islington', *Journal of Planning and Environment Law*, February: 121–6.

JURUE (1977) *Planning and land availability*. Report to the Department of the Environment. Joint Unit for Research in the Urban Environment, Birmingham: University of Aston.

Kavanagh, D. (1987) *Thatcherism and British Politics*, Oxford: Oxford University Press.

Kavanagh, D. (1990) *British Politics: Continuity and Change*, Oxford: Oxford University Press.

Kavanagh, D. (1994) 'A Major agenda?', in D. Kavanagh and A. Seldon (eds), *The Major Effect*, London: Macmillan.

Kavanagh, D. and Seldon, A. (eds) (1989) *The Thatcher Effect*, Oxford: Oxford University Press.

Kavanagh, D. and Seldon, A. (eds) (1994) *The Major Effect*, London: Macmillan.

Kearns, G. and Philo, C. (eds) (1993) *Selling Places: the City as Cultural Capital, Past and Present*, Oxford: Pergamon.

Keating, M. (1991) *Comparative Urban Politics*, London: Edward Elgar.

Kennet, Lord W. (1991) Letter to the Editor, *The Times*, 23 April.

Kent County Council (1993) *Kent Structure Plan Third Review Technical Working Paper 1/93: Strategic Environmental Appraisal of Policies*, Maidstone: Kent County Council.

Kent County Council (1994) *Kent Structure Plan Third Review Technical Working Paper 6/94: Kent Air Quality Model*, Maidstone: Kent County Council.

King, D.S. (1987) *The New Right: Politics, Markets and Citizenship*, London: Macmillan.

Kramer, L. (1995) 'Recent developments in EC environmental law', paper presented at a *Conference on the Impact of EC Environmental Law in the United Kingdom*, Centre for the Law of the European Union, London: University College.

Lancashire County Council (1994) *Greening the Red Rose County*, Deposit Version, Preston: LCC.

Larkham, P.J. (1991) 'The concept of delay in development control', *Planning Outlook*, 33: 101–7.

Larkham, P.J. (1994) 'Conservation areas and plan-led planning: how far can we go?', *Journal of Planning and Environment Law*, January: 8–13.

Larkham, P.J. (1995) 'Heritage as planned and conserved', in D.T. Herbert (ed.), *Heritage, Tourism and Society*, London: Mansell.

Larkham, P.J. (1996) 'Designating conservation areas: patterns in time and space', *Journal of Urban Design*, 1: 321–32.

Larkham, P.J. (1997) 'Conservation areas: ideal and reality reviewed', *Transactions of the Ancient Monuments Society*, 41.

Larkham, P.J. and Chapman, D.W. (1996) 'Article 4 Directions and development control: planning myths, present uses and future possibilities', *Journal of Environmental Planning and Management* 39: 5–19.

Larkham, P.J. and Jones, A.N. (1993) 'The character of conservation areas in Great Britain', *Town Planning Review*, 64: 395–413.

Lawson, N. (1992) *The View from Number 11*, London: Corgi.

Lean, G. (1995) 'Batty Redwood wants to Privatise Snowdonia', *Independent on Sunday*, 22 January.

Lean, G. (1996) 'Where have all the woods gone?', *Independent on Sunday*, 16 June.

Leedham, R. (1993) 'Feud's corner: Jocelyn Stevens vs the world', *The Times*, 18 July.

Le Grand, J. (1991) *The theory of government failure*, Studies in Decentralisation and Quasi-Markets 5, School for Advanced Urban Studies, University of Bristol.

Leicester City Council (1996) *City of Leicester Local Plan: Sustainability Appraisal*, Leicester: Leicester City Council.

Levitas, R. (ed.) (1986) *The Ideology of the New Right*, Cambridge: Polity Press.

Lipsky, M. (1978) *Street Level Bureaucracy*, New York: Russell Sage.

Lloyd, M.G. (1987) 'Simplified planning zones – the privatisation of land use controls in the UK', *Land Use Policy*, January.

Lloyd, M.G. (1996) 'Viewpoint: local government reorganisation and the strategic planning lottery in Scotland', *Town Planning Review*, 67, 3: v-viii.

Loftman, P. and Nevin, B. (1992) *Urban regeneration and social equity: a case study of Birmingham 1986–1992*, Research Paper 8, Birmingham: Faculty of the Built Environment, University of Central England.

Logan, J. and Molotch, H. (1987) *Urban Fortunes*, Berkeley, Calif.: University of California Press.

London First (1992) *London First: A World Class Capital*, London: London First.

Long, J. (1995) 'Unaccountable delay on route to plan adoption', *Planning*, 1127.

Lowe, P. and Flynn, A. (1989) 'Environmental planning and the Thatcher government', *ECOS*, 10, 4: 22–9.

Lowenthal, D. (1985) *The Past is a Foreign Country*, Cambridge: Cambridge University Press.

LPAC (1994) 'Advice on strategic planning guidance for London', London Planning Advisory Committee.

McAuslan, P. (1980) *The Ideologies of Planning Law*, Oxford: Pergamon.

McCormick, J. (1991) *British Politics and the Environment*, London: Earthscan Publications.

McDonic, G. (1988) 'Environmental impact assessment in planning practice: will it work?', in M. Clark and J. Herington (eds), *The Role of Environmental Impact Assessment in the Planning Process*, London: Mansell.

Macfarlane, R. (1993) *Community Involvement in City Challenge*, London: NCVO.

MacGregor, B. and Ross, A. (1995) 'Master or servant? The changing role of the development plan in the British planning system', *Town Planning Review*, 66, 1: 41–59.

MacGregor, D., Landridge, R., Adley, J. and Chapman, B. (1987) 'New firms and high technology industry in Berkshire', Department of Land Management and Development, University of Reading Working Paper.

McLaren, D. (1996) 'Achieving sustainability through the concept of "environmental space": a trans-European project' *European Environment*, 6: 69–76.

McLaren, D. and Adams, M. (1989) *Environmental Charter for Local Government*, London: Friends of the Earth.

McLaren, D. and Bosworth, T. (1994) *Planning for the Planet: Sustainable Development Policies for Local and Strategic Plans*, London: Friends of the Earth.

Maclennan, D. (1994) *Housing Policy for a Competitive Economy*, York: Joseph Rowntree Foundation.

Mark-Lawson, J. and Warde, A. (1987) 'Industrial restructuring and the transformation of a local political environment: a case study of Lancaster', Lancaster Regionalism Group Working Paper No. 33, University of Lancaster.

Marr, A. (1995) *Ruling Britannia*, London: Michael Joseph.

Marsden, T. and Murdoch, J. (1994) *Reconstituting Rurality*, London: UCL Press.

Marsden, T., Murdoch, J., Lowe, P., Munton, R. and Flynn, A. (1993) *Constructing the Countryside*, London: UCL Press.

Marsh, D. (1991) 'Privatisation under Mrs Thatcher', *Public Administration*, 69: 420–43.

Marsh, D. (1992) 'Industrial relations', in D. Marsh and R.A.W. Rhodes (eds), *Implementing Thatcherite policy. Audit of an Era*, Buckingham: Open University Press.

Marsh, D. and Rhodes, R.A.W. (1992) *Implementing Thatcherite Policies: Audit of an Era*, Buckingham: Open University Press.

Marshall, T. (1994) 'British planning and the new environmentalism', *Planning Practice and Research*, 9, 1: 21–30.

Marshall, T. (1994a) 'Dimensions of sustainable development and scales of policy-making', paper to ECPR Green Politics Standing Group, Crete, 21–23 October.

Meegan, R. (1993) 'Urban development corporations, urban entrepreneurialism and locality', in R. Imrie and H. Thomas (eds), *British Urban Policy and the Urban Development Corporations*, London: Paul Chapman.

Merrett, S. (1994) 'Ticks and crosses: strategic environmental assessment and the Kent structure plan', *Planning Practice and Research*, 9, 2: 147–50.

Merton, R.K. (1957) *Social Theory and Social Structure*, Glencoe, Ill.: Free Press.

Miller, C. (1994) 'Planning, pollution and risk', in *Town Planning Review*, 65, 2: 127–41.

Millichap, D. (1989) 'Conservation areas – Steinberg and after', *Journal of Planning and Environment Law*, April: 233–40.

Millichap, D. (1989a) 'Conservation areas and Steinberg – the Inspectorate's response', *Journal of Planning and Environment Law*, July: 499–504.

Millichap, D. (1993) 'Sustainability: a long-established concern of planning', *Journal of Planning and Environment Law*, 1111–19.

Ministry of Agriculture, Fisheries and Food CAP Review Group (1995) *European Agriculture: The case for radical reform*, London: Ministry of Agriculture, Fisheries and Food.

Ministry of Agriculture, Fisheries and Food CAP Review Group (1995a) *European Agriculture: The case for radical reform – working papers*, London: Ministry of Agriculture, Fisheries and Food.

Ministry of Agriculture, Fisheries and Food, Department of Agriculture Northern Ireland, Scottish Office Agriculture and Fisheries Department and Welsh Office Agriculture Department (1991) *Our Farming Future*, London: Ministry of Agriculture, Fisheries and Food.

Molotch, H. (1976) 'The city as a growth machine', *American Journal of Sociology*, 82, 2.

Monk, S. and Whitehead, C. (1996) 'Land supply and housing: a case study', *Housing Studies*, 11: 3, 407–23.

Monk, S., Pearce, B. and Whitehead, C. (1991) *Planning, land supply and house prices: a literature review*. Monograph 21, Department of Land Economy, University of Cambridge.

Montgomery, J. and Thornley, A. (1990) *Radical Planning Initiatives*, Aldershot: Gower.

Morris, J. (1995) 'Indicators of local sustainability', *Town and Country Planning* 64, 4: 113–16.

Morton, D.M. (1991) 'Conservation areas: has saturation point been reached?', *The Planner*, 77: 5–8.

Morton, D.M. and Ayers, J.H. (1993) 'Conservation areas in an era of plan led planning', *Journal of Planning and Environment Law*, March: 211–13.

National Rivers Authority (1994) *Guidance Notes for Local Planning Authorities on the Methods of Protecting the Water Environment through Development Plans*, Bristol: National Rivers Authority.

National Rivers Authority Thames Region (1995) *Thames 21: A Planning Perspective and a Sustainable Strategy for the Thames Region*, Reading: National Rivers Authority Thames Region.

NCVO (1993) *Community Involvement in City Challenge: a policy report and good practice guide*, London: National Council for Voluntary Organizations.

Newby, H. (1980) *Green and Pleasant Land*, London: Hutchinson.

Newby, H. (1990) 'Ecology, amenity and society: social science and environmental change', *Town Planning Review*, 61, 1: 3–13.

Newcastle upon Tyne City Council (1992) *Energy and the Urban Environment: Strategy for a Major Urban Centre*, Newcastle: Newcastle upon Tyne City Council.

Newman, P. and Thornley, A. (1997) 'Fragmentation and centralisation in the governance of London: influencing the urban policy and planning agenda', *Urban Studies*, 34, 7.

Newport Borough Council (n.d.), *Development and building control: a charter*, Newport: NBC.

Niskanen, W.A. (1971) *Bureaucracy and Representative Government*, New York: Aldine-Atherton.

North, R. (1995) *Life on a Modern Planet: a Manifesto for Progress*, Manchester: Manchester University Press.

Norton, P. and Aughey, A. (1981) *Conservatives and Conservatism*, London: Temple Smith.

Nuffield Foundation (1986) *Town and Country Planning: The Report of a Committee of Inquiry Appointed by the Nuffield Foundation*, London: Nuffield Foundation.

Oatley, N. and Lambert, C. (1995) 'Evaluating competitive urban policy: the city challenge initiative', in R. Hambleton and H. Thomas (eds), *Urban Policy Evaluation*, London: Paul Chapman.

O'Riordan, T. (1989) 'Nature conservation under Thatcherism: the legacy and the prospect', *ECOS*, 10, 4: 4–8.

O'Riordan, T. (1990) 'One and a half cheers for Chris Patten', *ECOS*, 11, 4: 2–7.

Owens, S. (1991) *Energy-Conscious Planning: The Case for Action*, London: CPRE.

Owens, S. (1994) 'Land, limits and sustainability: a conceptual framework and some dilemmas for the planning system', *Trans. Inst. British Geographers*, 19, 4: 439–56.

Oxfordshire County Council (1996) *Oxfordshire Structure Plan 2011: Environmental Appraisal of the Deposit Draft*, Oxford: Oxfordshire County Council Environmental Services.

Parkinson, M., Bianchini, F., Dawson, J., Evans, R. and Harding, A. (1992) *Urbanisation and the Function of Cities in the European Community*, Liverpool: European Institute of Urban Affairs, John Moores University.

Patten, C. (1990) 'The market and the environment', *Policy Studies*, 11: 4–10.

Peacock, A. (1984) 'Privatisation in perspective', *Three Banks Review*, 144: 3–25.

Pearce, D., Markandya, A. and Barbier, E. (1989) *Blueprint for a Green Economy*, London: Earthscan Publications.

Pearman, H. (1992) 'Suburban development: architecture', *The Sunday Times*, 6 September.

Petts, J. (1995) 'Waste management strategy development: a case study of community involvement and consensus-building in Hampshire', *Journal of Environmental Planning and Management*, 38, 4: 519–37.

Phillips, A.A.C. (1993) 'The Countryside Commission and the Thatcher years', in A. Gilg (ed.), *Progress in Rural Policy and Planning*, Vol. 3, London: Belhaven.

Pinfield, G. (1992) 'Strategic environmental assessment and land-use planning' *Project Appraisal*, 7, 3: 157–63.

Planning (1995) 'Planners to ditch detail after progress takes another knock', *Planning,* 1114: 1.

Planning Week (1996) 'Shakespeare's Stratford stages planning protest', *Planning Week*: 3.

Powys County Council (1994) *Powys County Structure Plan (Draft Replacement) Deposit Version*, Llandrindod Wells: Powys C.C.

Pressman, J. and Wildavsky, A. (1984) *Implementation*, Berkeley: University of California Press.

Pugh-Smith, J. and Samuels, J. (1996) 'Archaeology and planning: recent trends and potential conflicts', *Journal of Planning and Environment Law*, September: 707–24.

Punter, J.V. (1986) 'The contradictions of aesthetic control under the Conservatives', *Planning Practice and Research*, 1: 8–13.

Punter, J.V. (1990) *Design Control in Bristol 1940–1990*, Bristol: Redcliffe Press.

Punter, J.V. (1991) 'The long-term conservation programme in central Bristol 1977–1990', *Town Planning Review*, 62: 341–64.

Quinn, M. (1996) 'Central government planning policy', in M. Tewdwr-Jones (ed.), *British Planning Policy in Transition*. London: UCL Press.

Raban, J. (1989) *God, Man and Mrs Thatcher*, London: Chatto and Windus.

Ratcliffe, D (1989) 'The Nature Conservancy Council 1979–1989', *ECOS*, 10, 4: 9–15.

Ravenscroft, N. (1994) 'Leisure policy in the new Europe: the UK Department of National Heritage as a model of development and integration', *European Urban and Regional Studies* 1: 131–42.

Ravetz, A. (1980) *Remaking Cities*, London: Croom Helm.

Reade, E.J. (1987) *British Town and Country Planning*. Milton Keynes: Open University Press.

Reynolds, G. (n.d.) Work in progress on PhD on 'The influence of the pollution agencies on the planning system', Oxford: Oxford Brookes University School of Planning.

Rhodes, R.A.W. (1992) 'Implementing Thatcherite policies: an annotated bibliography', Essex Papers in Politics and Government, Colchester: University of Essex.

Rhodes, R.A.W. (1996) 'The new governance: governing without government', *Political Studies*, 44, 4.

Rhodes, R.A.W. and Marsh, D. (eds) (1992) *Policy Networks in British Government*, Oxford: Clarendon Press.

Riddell, P. (1983) *The Thatcher Government*, Oxford: Martin Robertson.

Riddell, P. (1991) *The Thatcher Era*, Oxford: Blackwell.

Ridley, N. (1992) *My Kind of Government*, London: Fontana Books.

Robertson, M. (ed.) (1993) 'Listed buildings: the national resurvey of England', *Transactions of the Ancient Monuments Society*, 37: 21–94.

Robinson, S. (1990) 'City under stress', *The Planner*, 76, 10.

Robson, B., Bradford. M., Deas I., Hall E. and Harrison, E. (1994) *Assessing the Impact of Urban Policy*, London: HMSO.

Rosen, B. (1993) 'Department gets pedantic about presumptions', *Planning*, September, 20–1.

Rowan-Robinson, J. and Lloyd, M.G. (1986) 'Lifting the burden of planning: a means or an end?', *Local Government Studies*, 12, 3: 51–63.

Rowan-Robinson, J., Ross, A. and Walton, W. (1995) 'Sustainable development and the development control process', *Town Planning Review*, 66, 3: 269–86.

Royal Town Planning Institute (1984) Letter to DoE.

Rumble, J. (1995) 'Environmental appraisal of Hertfordshire's structure plan', in R. Therivel (ed.), *Environmental Appraisal of Development Plans 2: 1992–1995*, Working Paper No. 160, Oxford: Oxford Brookes University School of Planning.

Rural Development Commission (1995) *Planning for Prosperity*, London: Rural Development Commission.

Rydin, Y. (1986) *Housing Land Policy*, Aldershot: Gower.

Rydin, Y. (1993) *The British Planning System: An Introduction*, London: Macmillan.

Sabatier, P. (1986) 'Top-down and bottom-up approaches to implementation research', *Journal of Public Policy*, 6.

Sabatier, P. and Mazmanian, D. (1979) 'The conditions of effective implementation: a guide to accomplishing policy objectives', *Policy Analysis*: Autumn.

Savage, S. and Robbins, L. (eds) (1990) *A Nation of Home Owners*, London: Unwin Hyman.
Schouten, F.F.J. (1995) 'Heritage as historical reality', in D.T. Herbert (ed.), *Heritage, Tourism and Society*, London: Mansell.
Scottish Office Development Department (1996) *Planning Advice Note 49: Local Plans*, Edinburgh: Scottish Office.
Seamark, D. (1996) 'European funding and environmental appraisal', *Town and Country Planning*, 65, 12: 340–1.
Shepherd, J. and Bibby, P. (1996) *Urbanization in England: Projections 1991–2016*, London: HMSO.
Simon, H.A. (1945) *Administrative Behavior*, Glencoe, Ill.: Free Press.
Skelcher, C. and Stewart, J. (1993) 'The appointed government of London', Paper prepared for the Association of London Authorities, November.
Slater, M., Marvin, S. and Newson, M. (1994) 'Land use planning and the water sector: a review of development plans and catchment management plans', *Town Planning Review*, 65, 4: 375–97.
Smith, D. (1989) 'Changing prospects in south Birmingham', in P. Cooke (ed.), *Localities*, London: Unwin Hyman.
Smith, M.J. (1990) *The Politics of Agricultural Support in Britain: the Development of an Agricultural Policy Community*, Dartmouth: Aldershot.
Solesbury, W. (1990) 'Property development and urban regeneration' in P. Healey and R. Nabarro (eds), *Land and Property Development in Changing Context*, Aldershot: Gower.
South Glamorgan County Council (1994) *Preparation of Energy and Emission Statements: Guidance Notes for Local Authorities and Developers*, Cardiff: South Glamorgan County Council and Cardiff City Council.
Spencer, K., Taylor, A., Smith, B., Mawson, J., Flynn, N. and Batley, R. (1986) *Crisis in the Heartland: a Study of the West Midlands*, Oxford: Clarendon Press.
Stamp, G. (1991) 'Heritage thrown to the piranhas', *The Spectator*, 3 August: 18–19.
Stevenson, D. (1994) 'Partnerships and the new sub-regions needed for resource bidding', in M. Edwards and J. Ryser (eds), *Bidding for the Single Regeneration Budget Workshop Proceedings*, London: UCL Press.
Stoker, G. (1995) 'Regime theory in urban politics', in D. Judge, G. Stoker and H. Wolman (eds), *Theories of Urban Politics*, London: Sage.
Stoker, G. and Mossberger, K. (1994) 'Urban regime theory in comparative perspective', *Environment and Planning C*, 12, 2.
Stone, C. (1989) *Regime Politics: Governing Atlanta 1946–88*, Lawrence: University Press of Kansas.
Stone, C., Orr, M. and Imbroscio, D. (1991) 'The reshaping of urban leadership in US cities: a regime analysis', in M. Gottdiener and C. Pickvance (eds), *Urban Life in Transition*, London: Sage.
Strathclyde Regional Council (1995) *Strathclyde Sustainability Indicators*, Glasgow: Strathclyde Regional Council.
Suddards, R.W. (1988) 'Listed buildings – have we listed too far?', *Journal of Planning and Environment Law*, August: 523–8.
Suttie, P. (1994) 'Going public', *Planning Week*, 24 February.
Taussik, J. (1992) 'Development Plans: implications for enhanced status', *Estates Gazette*, 9235, Sept 5th.

Tewdwr-Jones, M. (1994) 'The development plan in policy implementation', *Environment and Planning C: Government and Policy*, 12, 2: 145–63.

Tewdwr-Jones, M. (1995) 'Development control and the legitimacy of planning decisions', *Town Planning Review*, 66, 2: 163–81.

Tewdwr-Jones, M. (1997) 'Plans, policies and intergovernmental relations: assessing the role of national planning guidance in England and Wales', *Urban Studies*, 34, 1: 141–62.

Tewdwr-Jones, M. (forthcoming) 'Certainty and discretion in post-Thatcherite spatial planning in the UK', copies available from author.

Therivel, R. (ed.) (1995) *Environmental Appraisal of Development Plans 2: 1992–1995* Working Paper No. 160, Oxford: Oxford Brookes School of Planning.

Therivel, R. (1996) 'Environmental appraisal of development plans: status in late 1995', *Report*, 14: 14–16.

Therivel, R., Wilson, E., Thompson, S., Heaney, D. and Pritchard, D. (1992) *Strategic Environmental Assessment*, London: Earthscan Publications in association with RSPB.

Thomas, H. and Imrie, R. (1993) 'Cardiff Bay and the project of modernization', in R. Imrie and H. Thomas (eds) *British Urban Policy and the Urban Development Corporations*, London: Paul Chapman Publishing.

Thomas, H. and Imrie, R. (1997) 'Urban Development Corporations and local governance in the UK', *TESG,* 88, 1: 53–64.

Thornley, A. (1981) *Thatcherism and Town Planning*, Planning Studies No. 12, London: Polytechnic of Central London.

Thornley, A. (1986) 'Planning in a cool climate: the effects of Thatcherism', *The Planner*, 74, 7: 17–19.

Thornley, A. (1991) *Urban Planning under Thatcherism: the Challenge of the Market.* London: Routledge.

Thornley, A. (1993) *Urban Planning under Thatcherism: the Challenge of the Market*, second edition, London: Routledge.

Thornley, A. (1996) 'Planning policy and the market', in M. Tewdwr-Jones (ed.), *British Planning Policy in Transition*, London: UCL Press.

Thornley, A. and Newman, P. (1996) 'International competition, urban governance and planning projects.

Tiesdell, S., Oc, T. and Heath, T. (1996) *Revitalizing Historic Urban Quarters*, Oxford: Architectural Press.

Tompkins, S. (1989) 'Forestry in crisis', *ECOS*, 10, 4: 16–21.

Tunbridge, J.E. (1994) 'Whose heritage: global problem, European nightmare', in G.J. Ashworth and P.J. Larkham (eds), *Building a New Heritage*, London: Routledge.

Tunbridge, J.E. (1995) *Dissonant Heritage*, Chichester: Wiley.

Turok, I. (1992) 'Property-led urban regeneration: panacea or placebo', *Environment and Planning A*, 24.

Tyldesley, D. and Associates (1995) *Environmental Appraisal of the Gordon District Consultative Draft District Plan*, Hucknall, Nottingham: David Tyldelsey and Associates.

Tym, R. and Partners (1982) 'EZ Monitoring Report – Year One', London: Roger Tym and Partners.

Tym, R. and Partners (1983) 'EZ Monitoring Report – Year Two', London: Roger Tym and Partners.
Tym, R. and Partners (1984) 'EZ Monitoring Report – Year Three', London: Roger Tym and Partners.
Tym, R. and Partners (1990) *Housing Land Availability*, London: HMSO.
Vale of White Horse District Council (1995) *Vale of White Horse Local Plan Deposit Draft Written Statement*, Abingdon: Vale of White Horse District Council.
van Meter, D. and van Horn, C. (1975) 'The policy implementation process: a conceptual framework', *Administration and Society*, 6: 4.
Walker, B. (1981) *Welfare Economics and Urban Problems*, London: Hutchinson.
Ward, H. and Samways, D. (1992) 'Environmental policy', in Marsh, D and Rhodes, R.A.W. (eds), *Implementing Thatcherite Policy. Audit of an Era*, Buckingham: Open University Press.
Watkins, C. and Winter, M. (1988) *Superb Conversions? Farm diversification – the Farm Building Experience*, London: Council for the Protection of Rural England.
Watson, I. (1991) 'Demolition upheld in No. 1 Poultry decision', *Chartered Surveyor Weekly* 28 March: 73.
Weatherhead, P. (1994) 'Burning issues in a policy vacuum', *Planning Week*, 2, 14: 12–13.
Webster, B. and Lavers, A. (1991) 'The effectiveness of public local inquiries as a vehicle for public participation in the plan making process: a case study of the Barnet Unitary Development Plan Inquiry', *Journal of Planning and Environmental Law,* 803–13.
Weideger, P. (1994) *Gilding the Acorn: behind the Facade of the National Trust*, New York: Simon and Schuster.
Welbank, M. (1992) 'A green agenda for planning', *The Planner*, 78, 3: 8–9.
Welbank, M. (1993) 'Towards sustainable development', RSPB National Planners' Conference, *The New Age of Environmental Planning: Delivering Sustainable Land Use*, Sandy: RSPB.
Welsh Office (1993) *Development Control: A guide to good practice*, Cardiff: Welsh Office.
Welsh Office (1996) *Planning Guidance (Wales): Planning Policy*, Cardiff: HMSO.
West, B. (1988) 'The making of the English working past: a critical view of the Ironbridge Gorge Museum', in R. Lumley (ed.), *The Museum Time Machine*, London: Routledge.
West Sussex County Council (1995) *West Sussex Structure Plan Third Review Background Paper: The Environmental Capacity of West Sussex*, Chichester: West Sussex County Council.
Weston, J. and Hudson, M. (1995) 'Planning and risk assessment', *Environmental Policy and Practice*, 4, 4: 189–92.
Wilensky, H.L. (1964) 'The professionalisation of everyone', *American Journal of Sociology*, 70.
Willets, D. (1992) *Modern Conservatism*, Harmondsworth: Penguin.
Willis, R. (1988) 'New Zealand', in P. Cloke (ed.), *Policies and Plans for Rural People*, London: Unwin Hyman.
Wilson, E. (1993) 'Strategic environmental assessment: evaluating the impact of European plans, policies and programmes', *European Environment*, 3, 2: 2–6.
Wilson, E. (1994) 'Introduction to the issues', in E. Wilson (ed.), *Issues in the*

Environmental Appraisal of Development Plans, Working Paper No. 153, Oxford: Oxford Brookes University School of Planning.

Wilson, E. (1996) 'The precautionary principle and the compact city', in M. Jenks, E. Burton, and K. Williams, *The Compact City: A Sustainable Urban Form?*, London: E & FN Spon.

Wilson, E. and Raemaekers, J. (1992) *Index to Local Authority Green Plans*, 2nd edition, Research Paper No. 44, Edinburgh: Edinburgh College of Art/Heriot-Watt University.

Wistow, G. (1992) 'The national health service', in D. Marsh and R.A.W. Rhodes (eds), *Implementing Thatcherite Policy. Audit of an Era*, Open University Press, Buckingham.

Wolff, C. (1988) *Markets or Governments: Choosing between Imperfect Alternatives*, Cambridge, Mass.: MIT Press

Wood, C. (1988) 'The genesis and implementation of environmental impact assessment in Europe', in M. Clark and J. Herington (eds), *The Role of Environmental Impact Assessment in the Planning Process*, London: Mansell.

Wright, P. (1987) 'Why the blight is stark enough', *The Guardian*, 1 August: 17.

Young, H. (1994) 'The Prime Minister', in D. Kavanagh and A. Seldon (eds), *The Major Effect*, London: Macmillan.

INDEX